THE
SCIENCE OF
JAMES BOND

THE
SCIENCE OF
JAMES BOND

FROM BULLETS TO BOWLER HATS
TO BOAT JUMPS, THE REAL TECHNOLOGY
BEHIND 007'S FABULOUS FILMS

Lois H. Gresh

Robert Weinberg

WILEY

John Wiley & Sons, Inc.

Published by John Wiley & Sons, Inc., Hoboken, New Jersey
Published simultaneously in Canada

Design and composition by Navta Associates, Inc.

For general information about our other products and services, please contact our Customer Care Department within the United States at (800) 762-2974, outside the United States at (317) 572-3993 or fax (317) 572-4002.

Wiley also publishes its books in a variety of electronic formats. Some content that appears in print may not be available in electronic books. For more information about Wiley products, visit our web site at www.wiley.com.

Library of Congress Cataloging-in-Publication Data:
Gresh, Lois H.
    The science of James Bond / Lois H. Gresh, Robert Weinberg.
        p. cm.
    Includes bibliographical references and index.
    ISBN-13 978-0-471-66195-5 (pbk.)
    ISBN-10 0-471-66195-3 (pbk.)
1. Fleming, Ian, 1908–1964—Characters—James Bond. 2. Literature and science—Great Britain—History—20th century. 3. Fleming, Ian, 1908–1964—Knowledge—Science. 4. Spy stories, English—History and criticism. 5. Bond, James (Fictitious character) 6. Science in literature. I. Weinberg, Robert E. II. Title
    PR6056.L4Z66 2006
    823'.914—dc22   2005013194
Printed in the United States of America

10 9 8 7 6 5 4 3 2 1

To Sean Connery and Ursula Andress,
who started it all. To Blofeld and Jaws.
And, most definitely, to Q.

Contents

Foreword

By Raymond Benson

I was nine years old when I made my father take me to see my first James Bond film, *Goldfinger*. The time was early 1965, probably January, because the film was on its second run at a drive-in theater in Odessa, Texas, where my family was living. My next-door friends' mom had been playing the soundtrack on their stereo one day, and I thought the music sounded really cool. I looked at the pictures on the record sleeve and was impressed by the sexy women and the suave guy, whose name was apparently "James Bond." Some of my friends had seen the movie and told me I shouldn't miss it. So my dad and I went to the drive-in, and it forever changed the way I looked at the world.

The James Bond films were the *Star Wars* of the 1960s. They were the blockbusters, the movies everyone stood in line around the block to see. There was nothing else like them until they began to be imitated, usually poorly. Bond was the king of the action/adventure genre until the series shot itself in the foot at some point during the seventies, when the movies became action/*comedies* instead of films capable of generating real suspense.

I really didn't care, though. I remained a Bond fan as I went through high school and onward to college. I kept the faith after receiving a degree and moving to New York City to pursue a career in the arts. My fascination with the character (and especially Ian Fleming's original novels) ultimately culminated in writing *The James Bond Bedside Companion*, a true labor of love. One might have thought that the book, published in 1984, would have satisfied my

Bond addiction—but no, it only made me more obsessive. A little over a decade later I found myself penning official 007 novels for Ian Fleming's family and company. Who would have thought?

Exactly what was it that first hooked me into Bondmania? I think it was probably three things: the music, the ultra-coolness of Sean Connery, and the fantastic hardware. I doubt that there was anyone who saw *Goldfinger* at the time of its release who wasn't wowed by that Aston Martin. When Bond pressed the red button ("Whatever you do, don't touch it!" warned Q) and the Korean guard went flying, that was it; the filmmakers had me, and everyone else, in the palms of their hands.

Of course, I was nine years old. I didn't care that an ejector seat like that would have been impossible to achieve, as the authors of this book point out. I *wanted* to believe in it. And I accepted all of Bond's hardware as real in the first three films (I caught up with the earlier two on a double-bill re-release in the summer of 1965).

When *Thunderball* came out at the end of 1965, I was still at an impressionable age, but I distinctly remember being struck by the bit when the Aston Martin spurted liquid out of its exhaust pipes with the force of a fireman's hose for nearly a minute. I turned to my parents (my mom had accompanied my dad and me to this one) and asked, "Where did all that water come from?" But I didn't really care. It was *Bond*!

Then came *You Only Live Twice*, in which it seemed that all the vehicles and rooms were equipped with closed-circuit televisions and monitors. In the pre-credits sequence we see a space capsule being swallowed up by SPECTRE's rocket. The guys down at NASA could watch the same scene on their monitors, just like *we* saw it. Even at age twelve, I turned to my father again (it had become a tradition to see Bond films together) and asked, "Where's the camera? Is there another spaceship out there?" But again, I really didn't care.

Alas, many of the gadgets and hardware became more and more outlandish and impossible. A Lotus Esprit that performed like a submarine? A gondola that turned into a hydrofoil? A submarine disguised as an ice floe? An Aston Martin that can turn *invisible*?

As the Bond films evolved and I grew older, I learned that in order to appreciate the pictures for what they were, one had to suspend disbelief to a certain degree. Sometimes that degree was huge. You went along for the ride and enjoyed it, or you stood up indignantly and shouted, "Oh, come on! No way!" I have to admit there were a few times when I did the latter.

From 1996 to 1997 I wrote my second Bond novel, *The Facts of Death*. For that book I wanted to put 007 in a new car. I thought about using the new Aston Martin but eventually settled on the Jaguar XK8. I contacted Jaguar in England and developed a relationship with one of the car's principal designers, a man with a lively sense of humor named Fergus Pollock. Totally game, Fergus helped me come up with the gadgets and weaponry that would go into *my* Bond car. He made several suggestions, many of which caused me to react with, "Oh, come on! No way!" But Fergus insisted that his ideas were plausible—if not today, then certainly tomorrow. These included a small flying "scout" that operated much like Batman's bat plane. It was stored beneath the chassis, and Bond could fly it by remote control, using a joystick. It seemed farfetched, but Fergus assured me it was actually in the planning stages at Jaguar, so I used it.

Another one of Fergus's ideas was color-changing pigment, allowing the Jaguar to be gray in one country and red in another—sort of a variation on the revolving license plates used in *Goldfinger*. Fergus claimed this could be achieved by electrically charged pigments in the paint, so I used it. In fact, this may be the theory behind the invisible car in *Die Another Day*.

There were other spiffy items such as the hologram projector, which could fool the baddies into believing that another car was heading toward them in their lane, forcing them to turn off the road and crash. And I had the standard features such as machine guns, oil slicks, smoke screens, and GPS homing devices.

I received such positive feedback from fans that I even brought the XK8 back in my third novel, *High Time to Kill*. But just as Ian Fleming himself received the dodgy letter or two from fans complaining about this or that, I also received my share of criticism for creating "impossible" gadgets. One fan took me to task on the

Internet, complaining that the things I had in the Jaguar could never be done, and the car was more fitting for a DC or Marvel comics superhero. Never mind that I had one of the Jaguar's designers helping me with the weaponry. Never mind that the stuff was just plain *fun* and would make terrific action sequences in Bond films, should the stories ever be adapted for the screen. My gadgets were simply created in the spirit of what had become the norm and what was *expected* to be in a Bond story.

Was my Jaguar XK8 cool? I thought so. Was the car a little on the outlandish side? Yes, absolutely. I didn't care that it blended today's fact with tomorrow's fiction because *this was Bond*!

The Science of James Bond is for the skeptics out there who want their science explained. It separates the fact from fiction and the plausible from the outlandish. The authors have done a splendid job in pointing out what can really be done and simultaneously presenting a historical perspective on the hardware of the spy business. With convincing research and not just a little humor, authors Ms. Gresh and Mr. Weinberg have given Bond fans an entertaining user's manual for 007's many tricks of the trade.

And at the end of it I found that I cared. A lot.

Raymond Benson was the official author of James Bond continuation novels between 1996 and 2002. He wrote six original 007 novels, three film novelizations, and three short stories during his tenure. He is also the author of the original suspense novels *Sweetie's Diamonds*, *Face Blind*, and *Evil Hours*. Writing as David Michaels, he is also the author of the best-selling *Tom Clancy's Splinter Cell* and *Tom Clancy's Splinter Cell—Operation Barracuda*. Go to www.raymondbenson.com for more information.

Introduction
Bond, James Bond

The name's Bond, James Bond.
—AGENT 007 TO DR. NO, *DR. NO*

A medium-dry martini, lemon peel, shaken, not stirred.
—DR. NO TO AGENT 007, *DR. NO*

Everyone knows that Agent 007 is James Bond, the most famous secret agent of all time. Of course, we also know that Bond is a *fictional* secret agent created by Ian Fleming in the 1950s. Bond appeared in fourteen Fleming novels, the first of which was *Casino Royale* in 1953. He later appeared in twenty authorized films, the first being *Dr. No* in 1962. We're not counting the 1954 television version of *Casino Royale* starring Barry Nelson, the Woody Allen comedy film based on the same book, or *Never Say Never Again*, the 1983 remake of *Thunderball*. At the writing of this book, the twenty-first feature film starring James Bond was in the works.

James Bond is an elite spy with a top-secret 00 prefix, giving him a license to kill. He is a master of disguise, a master of superscience gadgets, and a master of seduction. He drinks martinis, "shaken, not stirred"—although in *Dr. No* he also insisted on a lemon peel—while wearing $6,000 suits and gambling away tens of thousands of dollars. He drives $200,000 cars equipped with rocket launchers, tire slashers, signaling devices, and submarine and airplane capabilities. He fights superevil villains beneath the earth, in the ocean, all over the planet, and even in outer space. And somehow, during all this mayhem, he manages to bed at least several beautiful women per adventure.

While many people dream about Bond's sartorial, gambling, and sexual skills, they also dream about his gadgets: the Walther PPK gun, golden guns, ray guns, and rocket-firing cigarettes; the spy shaving kit; poisonous shoes; homing device buttons; the Geiger counter wristwatch, the television wristwatch, and the buzz saw wristwatch; decoders; voice synthesizers; cameras; decapitating tea trays; killer hookahs; moonbuggies; underwater propulsion devices; minijets; and cars, cars, *cars*. And that's just a glimpse of the vast array of superscience gadgets issued by Q Division to Agent 007.

This book is about the gadgets, the science, the cars, and the technology. Other writers have covered James Bond's sartorial and sexual skills. We find the gadgets from Q Division far more interesting. How realistic are James Bond's adventures and all the equipment that goes with them? How believable is the villain Jaws, who chews through metal? Just how easy is it to crack into all those top-secret facilities to save the world? Is James Bond rooted in science, or are his adventures and the technology that fills them science fiction? Read on for the answers to these and other questions surrounding the science of Mr. Bond.

Uncovering the Origins of Mr. Bond

Spies and Science

I never joke about my work!
—Q TO 007, GOLDFINGER

Before delving into the main topic of our book, we need to take a short but very necessary detour. Let's look at James Bond's place in the world of real spies and peek at some of his fictional counterparts. Our objective, or mission, if you prefer, is to discover why James Bond is so popular. One theory is that his persona excites men and fascinates women. Another theory is that we're visually and mentally stunned by his toys—the cars, the gadgets, and the spy paraphernalia. We suspect both viewpoints are equally true. Of course, what appealed to fans forty years ago is much different from what appeals to them today.

If we study the history of spies in real life and in fiction, we see that spying has evolved from deception and espionage to the use of high-tech gadgets and computer hacking. The James Bond movies reflect the times during which they were produced, and they've evolved along similar lines in the real-life spy world. It's important to realize that Bond remains popular because he *has* changed with the times. The James Bond movies of today, with the superweapons and invisible cars, are vastly different from the Bond movies of the 1960s, when Bond used a chartered fishing boat to investigate Dr.

No and British intelligence was anxious to obtain a Russian Enigma decoder device. James Bond is a reflection of our desires and tastes, and therefore he always remains up-to-date.

A Short History of Spying

Spying began when humankind began. The earliest spies had no gadgets, computers, or technology. Rather, they merged into foreign societies and then reported back to their homelands about life there, secrets uncovered, and vulnerabilities. According to the Bible, Joshua sent spies into the land of Canaan to infiltrate the walled city of Jericho and learn its weaknesses. Twenty-five hundred years ago, Sun Tzu of China sent spies into other countries to learn their secrets; his objective was to avoid war.

In ancient Greece, the Spartans transmitted encoded information between secret agents and generals. The first spy gadget, the enigma device of the time, may have been a baton wrapped with encoded papyrus or parchment. After the encoded message was written, the papyrus or parchment was unwrapped from the baton, and then a secret agent delivered the message. Upon delivery, the message could only be read by wrapping the papyrus or parchment around an identical baton.

The ancient Romans also encoded their messages. Later, Genghis Khan used spies to gather information and spread negativism among the enemy. The Khan's spies were the equivalent of the Nazis' infamous Fifth Column and of today's double agents.

During the fifteenth century, espionage was common in Europe. A secret police force called the Council of Ten employed double agents throughout the Continent.

Queen Elizabeth I had spies, who discovered a murder plot against her. Supporters of Mary Queen of Scots, led by Anthony Babington, were planning an assassination. The spies also learned that King Philip II of Spain was in on the plot. The plan collapsed because Babington's spy, Gilbert Gifford, was actually a double agent working for Sir Francis Walsingham, the head of Queen Elizabeth's secret agents. One of Walsingham's spies happened to be the playwright Christopher Marlowe, a contemporary of William Shakespeare. Had James Bond lived back then, he might have been a wannabe playwright, or perhaps a moderately successful poet; not quite the dashing figure of today. (As professional writers, we feel qualified to tell you that poets and playwrights are not as dashing as James Bond. Bob does not drive an Aston Martin, nor does he drink martinis, shaken, stirred, whipped, dipped, or otherwise concocted. Lois does not wear heavy makeup and evening gowns, nor does she have a pseudonym along the lines of, say, Pussy Galore.)

At any rate, due to espionage, Mary Queen of Scots was beheaded, and everyone who supported her was hanged (and drawn and quartered; plotting against an absolute monarch was risky business!).

During the same general time period, France also ran an espionage unit under Cardinal Richelieu. The cardinal's skill was formidable basically due to his ties to the Catholic Church, which operated a highly successful team of secret agents. The Cabinet Noir, established in 1620 by the cardinal and headed by the Catholic priest François le Clerc du Tremblay, was analogous to Bond's Directorate of Military Intelligence, Section 5 (MI5) or the United States' Central Intelligence Agency (CIA). They routinely intercepted and decoded spy messages and carefully analyzed world situations, hoping to thwart potential plots against France.

Now we'll swing through time into the eighteenth century. We've established that James Bond would have been a poet or a playwright—or possibly a Catholic priest—had he been a secret agent earlier than the eighteenth century.

In 1780, spies in George Washington's Culper Ring used invisible ink called Jay's sympathetic stain. A reagent was later used to reveal the writing. The messages were placed in boxes that were buried in a cow pasture. The arrival of fellow agents was signaled by hanging a woman's black slip on a clothesline. Benjamin Tallmadge, the head of Washington's secret service, also used three-digit codes (think 007) for his agents. In fact, during this time, one of the most famous double agents in history, General Benedict Arnold, was reporting American secrets to Britain. Had James Bond been part of the Culper Ring, he would have been found rowing along the shores of Long Island, seeking a fluttering black slip and a cow pasture. Rather than fighting Dr. No, he would have been fighting Benedict Arnold—possibly using spears hidden within his oars.

Fast-forward to World War I in 1915. The British professor R. V. Jones figured that the Germans were conducting weapons research in Peenemünde, Germany, on the Baltic coast. There was no real evidence that this was happening, but Jones noted that a lot of German gas and oil was allocated to Peenemünde; hence something important was going on there. Jones concluded that the rumors about weapons research were true. Again, the spy work involved analysis rather than fast cars, fast women, and high-tech gadgets.

Deception and espionage were also involved during World War I, when the British developed the Q-ship in response to the German deployment of the U-boat. (Possibly the reason why Q Division was a Q rather than an M, a Z, or an F.) Here we see a turning point in spy history: the use of technology.

The U-boats were sinking Allied ships that carried supplies and food across the Atlantic Ocean to Britain. The ships were traveling without protection, so it was easy for U-boats to surface and use their deck guns to sink the unarmed vessels.

To counter the U-boats, the British designed the Q-ship, which was manned by sailors posing as plain seamen and wearing ordinary clothing. The Q-ship was heavily armed with guns, and when the ship's crew saw a U-boat surface, they would hoist a warship flag and blast the U-boat with gunfire. It was only when the spy Jules Crawford Silber told the Germans that the Q-ships were real that the Germans learned how their U-boats were being destroyed. The Germans changed their U-boat tactics. Instead of rising to the surface, submarines remained underwater and sank British ships using torpedoes. Advanced technology triumphed over deception. In the end, the British and the Americans were forced to group their supply ships in convoys with naval escorts to elude the U-boats.

In World War II, espionage and technology soared. In 1939, the British captain Hector Boyes received a letter that offered him a way to obtain information about German science and technology. All Boyes had to do was get the British Broadcasting Corporation (BBC) to broadcast "Hallo, Hallo, hier ist London" rather than the usual "Hallo." Boyes arranged for the longer Hallo broadcast and awaited developments. He soon received seven pages of typed espionage material, along with part of a proximity fuse for German antiaircraft shells. The espionage material included information about Peenemünde research into radio-controlled glider bombs, bomber guidance systems, and early-warning radar systems. It also contained details about a Junkers 88 dive-bomber that the Germans were developing, and it described German research into advanced torpedoes.

Professor Jones, who now headed Britain's Air Ministry Intelligence unit, received all the material from Boyes. Jones requested help from British scientists, who told him that all the material was false, but of course Professor Jones disagreed with them. His remarkable insight led him to believe that the material was indeed real because the proximity fuse was real. Over time, Jones and his agents used the information to uncover German bombing guidance and radar systems and V1 and V2 missiles.

We have to remember how important this espionage information was to the safety of the world. In World War II, Adolf Hitler deployed the first V1 cruise missile and the first V2 ballistic missile. The V2 missiles nearly destroyed London and Antwerp.

Many years later, in the 1950s, the author of the espionage material surfaced. Hans Ferdinand Meyer had been working in Germany for Siemens Electronics. He feared a Nazi victory should the Germans develop and deploy secret weapons. He wanted to live in a peaceful Germany. Poor Meyer ended up in a concentration camp. His crime? Listening to the BBC.

In 1941, the Russians sent a superspy into Tokyo. Posing as a Nazi journalist, Richard Sorge was so highly regarded by the Nazis that the German minister for propaganda, Joseph Goebbels, attended a dinner in Sorge's honor. As a Russian secret agent, Sorge learned that Germany planned to attack Russia on June 20, 1941. He warned his country, but Joseph Stalin did not believe him.[1]

The Germans attacked Russia, of course. Then Sorge learned that Japan planned to attack the United States and England in early December, while simultaneously attacking Russia. He warned

Stalin, who believed him this time. Literally within days of delivering this vital information to Russia, Sorge was hanged by the Japanese for espionage.

Also operating out of Russia during World War II was the Red Orchestra espionage group. The Red Orchestra had spies in Belgium, Holland, Switzerland, and Germany. They uncovered many German secrets until 1941, when Germany traced secret messages to some houses in Brussels, Belgium. The Germans arrested the Russian spies and confiscated their spy equipment: invisible ink and coding books. Then in 1942, the Nazis discovered Johann Wenzel of the Red Orchestra. Wenzel was transmitting secret, encoded messages about German operations and research to the Russians. The Germans used Wenzel to give false information to the Russians until, nearly a year later, the Russians got wise to what was going on and shut down the Red Orchestra.

There are many other cases of espionage, deception, and technology during World War II and that general time period. The Spy Museum in Washington, D.C., has many examples of spy technology from that era, including:

- The Minox Riga camera, issued by Latvia to secret agents from 1937 to 1944. It measures only 1 inch by 2 inches.
- The Steineck ABC wristwatch camera, issued by Germany to its agents in 1949. It has a 1-inch-wide film disk and the ability to produce eight exposures.
- The heel knife, issued by the British Special Operations Executive to its agents from 1940 to 1945.
- The tear-gas pen, issued by the U.S. Central Intelligence Agency to its agents in 1948.
- The tobacco pistol pipe, issued by the British Special Forces to its agents from 1935 to 1945.
- The rectal concealment cyanide capsule, used in 1945 by Hermann Göring, the head of the World War II German Luftwaffe, to commit suicide. He stuck a rifle cartridge holding a cyanide capsule up his rectum and then pulled out the device while in a Nuremberg prison.

It's not a stretch to think of James Bond using any of the spy equipment from the World War II era, except perhaps the rectal concealment cyanide capsule, which might have offended his British sense of dignity a bit. But even in the 1960s and 1970s, Bond had camera wristwatches, tear-gas pens, and heel knives. If anything, some of his most popular gadgets were old-hat spy tools.

Bond in Print

James Bond's creator, Ian Fleming, was a British citizen who was born on May 28, 1908. His father died in World War I, and Fleming later served as a commander of Naval Intelligence during World War II. It was then that the seeds of James Bond, Agent 007, were planted. Fleming was the right-hand man to Admiral John Godfrey, one of Britain's top spies. Although Fleming did not operate in the field, traveling like Bond around the world on adventures, he did serve as an espionage analyst and was a close friend of Bill Donovan, a top man in the Office of Strategic Services (OSS) and later in the CIA.

When the war ended, Fleming needed to earn a living. Having grown up in a socially elite British family, having lured a married woman of high circles, Lady Ann Rothermere, away from her husband—she became pregnant, got a divorce, and then married Fleming—and having a strong need to prove to his new wife that he was capable of enabling her to flourish in high social circles, Fleming needed cash.

In 1952, he started working on the novel *Casino Royale*. He finished the book in four weeks, and James Bond was born.

Remember that in 1952 Britain was recuperating from World War II. The rationing was ending, as was the general postwar austerity. In the United States, people bought homes and cars, tried to climb social ladders, and focused on rebuilding life overall. People went to the movies to see glamour, flamboyance, and fantasy lifestyles. The same was true in Britain.

The creation of a character who was larger than life, whose lifestyle was every man's dream, was what the world wanted: a man

who wore dinner jackets to casinos, who smoked only the finest cigarettes, who drank only the finest liquors, who slept with only the
most beautiful women, who drove only the most exotic and expensive cars, who traveled throughout the world on amazing adventures, who had the audacity and ability to demand and get a
"medium-dry martini, lemon peel, shaken, not stirred,"[2] who was
charming, brilliant, famous, and feared. This man was irresistible to
the post–World War II audience, and he was probably the man that
Ian Fleming wanted to be.

As for Agent 007, we note that Fleming, as a Naval Intelligence
officer, analyzed and processed documents all day. These documents
actually had a prefix attached to them. That prefix was 00. It so happens that Fleming's father was a friend of Winston Churchill's.
Winston's ancestor John Churchill, the duke of Marlborough, gave
his spies a 00 code during the War of the Spanish Succession from
1701 to 1714. But where does the 7 in 007 come from? For that, we
turn to a spy for Queen Elizabeth I, Dr. John Dee. Dee traveled
extensively; he was brilliant and dashing. The queen sent Dee on
top-secret missions to thwart King Phillip II's plans. Dee sent secret
messages to the queen using a special code for himself. That code
was 007. Supposedly, the 00 stood for two eyes, indicating to the
queen that the message was "for your eyes only." The 7 was *the*
lucky number.

So when Fleming wrote *Casino Royale* in 1953, his Agent 007
was a figure of great fancy to cold war–era people worldwide. At
that time, people could believe that the Russians were capable of the
evil presented by SMERSH (a Soviet KGB organization devoted to
assassination and violence) and SPECTRE (Special Executive for
Counterintelligence, Terrorism, Revenge, and Extortion).

But was 007 the first spy in fiction? Absolutely not. In 1821, a
novel called *The Spy* was published. Its author was James Fenimore
Cooper, who also wrote *The Last of the Mohicans*. Cooper's book featured a British spy named Major Andre. A number of other spy novels were published throughout the nineteenth century, but most of
them were not big sellers. Historical novels were all the rage; spies
were not in fashion. Then in 1894, William Le Queux's *The Great*

War in England in 1897 thrilled thousands of British readers with its focus on the terrifying specter of German spies. Le Queux wrote numerous spy novels in the days before World War I and was perhaps the best-selling British author of the period.

The Riddle of the Sands (1903) by Robert Erskine Childers predicted a British war with Germany and called for increased British readiness. It was an extremely influential book; many experts consider it the first realistic spy novel. Winston Churchill credited it as a major reason for the British to establish naval bases at Invergordon, the Firth of Forth, and Scapa Flow. In an odd twist of fate, after fighting for the British in World War I, Childers became involved in the battle for Irish independence. A member of the Sinn Féin, Childers was considered to be the man behind the Republican terrorists' tactics. He was captured by Irish Free State soldiers in November 1922 and was executed in Dublin by firing squad. He was one of the few spy novelists ever to become a real spy.

In 1907, Joseph Conrad's *The Secret Agent* presented spies who were shady underworld characters. Then in 1915, Sir John Buchan's *The Thirty-nine Steps* was published to great acclaim in both England and the United States. In 1935, Alfred Hitchcock made it into an effective spy thriller starring Robert Donat and Madeleine Carroll. Buchan later served as the governor general of Canada. Somerset Maugham produced a number of spy novels, followed by Graham Greene much later. Greene's secret agents were a miserable lot and tended to drink too much whiskey.

In mainstream fiction, most spies were portrayed as ordinary men thrust into extraordinary circumstances. While they had a certain heroic attitude, most of them were not terribly brave, and several of them got their courage from a bottle. It was Ian Fleming's Bond, however, who changed the genre of spy fiction from the anti-hero alcoholic character to the daring, dashing hero who drinks to impress but never gets drunk.

After Bond came spies created by Len Deighton and John le Carré. Deighton's novels were very popular in the 1960s; his spy hero wore thick glasses and didn't seem to attract many women. Le Carré's Leamas in *The Spy Who Came In from the Cold* was also a loner, and his escapades seemed more real and cold than, say, Bond's shenanigans.

The most popular spy novelists of the last quarter of the twentieth century were Tom Clancy, Robert Ludlum, and Eric van Lustbader. Clancy's novels about superspy Jack Ryan were filled with extensive details about weapons and weapons systems, and they found an appreciative audience among armchair soldiers. Ludlum wrote numerous spy novels about world-shaking events usually involving gigantic conspiracies that could only be stopped by one man or one woman or sometimes neither. His plots twisted; the novels were page-turners to the end. Van Lustbader's novel *The Ninja*, more than any other modern novel, focused attention on the secret spy order of Japan and their seemingly more-than-human athletic and spy skills. It was Lustbader's early spy novels that focused the attention of spy readers from Russia to Japan.

Bond Hits the Movies

Fleming's Bond books were well received but not best sellers. Several of the hardcovers even had their names changed when reprinted in paperback in the United States. *Casino Royale* became *You Asked for It*, while *Moonraker* was retitled *Too Hot to Handle*. Fleming was one of numerous writers who were publishing espionage novels, and Bond was one of many postwar spies. By the late 1950s, when Fleming was working on *From Russia with Love*, he considered disposing of Bond forever. But he decided to continue with the series, which proved to be a very wise decision a few years later.

In the early 1960s, John F. Kennedy was both the president of the United States and a big fan of James Bond. In fact, in an interview it was reported that Kennedy's ninth favorite book was *From Russia with Love*. The president's interest in the Bond books helped push them onto the best-seller lists and made Fleming and James Bond famous. It was during that period that producers Albert R. "Cubby" Broccoli and Harry Saltzman bought the film rights to James Bond, and Bond was then transformed into a spy icon.

Before Kennedy was president, he had dinner with Fleming,

arranged by a friend of Fleming's who had given Kennedy a copy of *Casino Royale* in 1955. The two men discussed Fidel Castro over dinner. Years later at the White House, Kennedy arranged for a private showing of *Dr. No.*

During the 1950s and 1960s, real spies used many of the gadgets that Q Division gave to Bond. Here are some examples, again from the Spy Museum in Washington, D.C.:

- A hairbrush compartment for the Minox camera
- The Echo 8 cigarette lighter camera, created by Japan and used by U.S. intelligence
- The Tessina camera and cigarette case—a cigarette pack with a hole in its side to conceal a tiny camera lens—issued by the East German Stasi
- The Toychka camera in a necktie, issued in 1958 by the KGB to its agents
- A shoe with a heel transmitter, issued in the 1960s by the KGB to its agents
- A rectal toolkit, filled with escape tools, issued by the CIA to its agents in the 1960s
- A lipstick gun that supplied one 4.5 mm shot, issued by the KGB in 1965 to its agents
- A gun in a cigarette case that delivered one shot from a .22 caliber gun, issued in the 1950s by the KGB to its agents
- A wristwatch microphone, issued by the United States in 1958 to its agents

It's clear that Bond continued to use devices that real spies were already using. His cool wristwatches, guns, and shoes were being issued by spy organizations around the globe. You could almost argue that James Bond was the spy fiction equivalent to John F. Kennedy himself: dashing, handsome, full of adventure and charisma, a lady's man, and stoked with plenty of power. Though it's a bit of a push to link the two this closely, it would not be surprising to discover that Fleming gave James Bond various Kennedy attributes along the way.

Modern Spies

Let's return to *From Russia with Love*. Fleming postulated that Bulgarian secret agents would handle assassinations and other dirty espionage work for the Russians. In 1978, the Bulgarian dissident Georgi Markov was killed by a ricin pellet fired from an umbrella. The Soviets admitted that the Bulgarian secret service was responsible for the assassination.

In the 1961 novel *Thunderball*, Fleming suggested that evil villains were stealing nuclear weapons and planning to destroy the world's largest cities unless the modern superpowers gave them $100 million. Such fears were actually quite true in the early 1960s, and they remain true to this day.

The threat of mutual assured destruction (MAD) grew after the cold war, and new horrors and technical wizardry arose: Ronald Reagan's 1980s Star Wars antimissile defense systems, which prompted the Soviets to gear up their own arms race; chemical and biological weapons; and suicide bombers. Saddam Hussein, for example, North Korea's Kim Jong-il, and Osama Bin Laden are world terrorists today who may press the nuclear, chemical, and/or biological buttons that destroy the world.

Today's espionage focuses on far more than exploding neckties, rectal concealment cyanide capsules, and lipstick guns. The world is a very unsafe place. It was a true nightmare during World War II, and that nightmare has only been exacerbated by global terrorism. Today's real spies focus on airborne and satellite intelligence and electronic and computer communications, and the world of James Bond has also evolved—at least his gadgets have evolved a bit, as shown in this chronological list of the Bond films detailing some of the more interesting devices used in the films:

- *Dr. No* (1962): a Walther PPK 7.65 mm firearm. In our opinion, not very high-tech.
- *From Russia with Love* (1963): a camera recorder. Again, not very high-tech.
- *Goldfinger* (1964): another stunning Aston Martin DB5. Our opinion: a cool movie, but not many Q gadgets.

- *Thunderball* (1965): an infrared film camera, compressed-air missiles, a Geiger-counter wristwatch, and a stunning Aston Martin DB5. Bond's technology is definitely increasing now, along with the times.
- *You Only Live Twice* (1967): a cigarette with a deadly dart, homing missiles, and an autogyro. On the general level of *Thunderball's* science.
- *On Her Majesty's Secret Service* (1969): a computerized copy machine and a safecracking device. Computers are in this one, but otherwise Bond is still somewhat low-tech.
- *Diamonds Are Forever* (1971): a voice box that imitates other people, a casino cheating device, and artificial fingerprints. More computerization in James Bond's bag of spy tools.
- *Live and Let Die* (1973): a circular-saw wristwatch. Fairly low-tech for the times.
- *The Man with the Golden Gun* (1974): a camera with rockets and a button homing device. Again, fairly low-tech for the time; this one's more like *Fantasy Island* with spies.
- *The Spy Who Loved Me* (1977): a cigarette-case microfilm viewer, a wristwatch with ticker tape, and a Lotus car–submarine. Very cool underwater technology.
- *Moonraker* (1979): a cigarette-lighter camera, a cigarette-case safecracking device, a wristwatch with a bomb detonator, speedboats, and a hovercraft. Lots of strange technology in this space-age Bond film. This one reminds us of *Star Wars*.
- *For Your Eyes Only* (1981): a binocular camera. This one is more glam-Bond; not much in the way of high-tech hijinks.
- *Octopussy* (1983): a television wristwatch, a video camera, a wristwatch homing device, a pen with metal-cutting acid, and a car rickshaw. A spy macho movie; the technology isn't as savvy as we'd like, given that the film came out in 1983, when computers were already widespread.
- *A View to a Kill* (1985): a robot surveillance device, a pen of burning words, a man's shaver that detects bugs, and a video camera that supplies identities of people using a centralized computer system (possibly a mainframe or a series of computers

linked together). A very high-tech Bond: seven stars out of ten on the Bond Gadget Scale.

- *The Living Daylights* (1987): a radio receiver pen, a key-ring stun-gas pistol, a key-ring bomb, and an amazing Aston Martin V8 Volante car. A techno-retro Bond: four stars out of ten on the Bond Gadget Scale.
- *Licence to Kill* (1989): an exploding alarm clock, a gun camera, an X-ray camera, and a toothpaste bomb. Sadly, not much science here.
- *GoldenEye* (1995): a silver tray X-ray document scanner, a grenade pen, a wristwatch that arms bombs and operates as a laser, and a leg-cast missile launcher. What can we say?
- *Tomorrow Never Dies* (1997): a wristwatch bomb detonator, a Sea-Vac underwater drill boat, a cell phone with a fingerprint scanner, and a very high-tech BMW 750 iL car. Cool tech in the car.
- *The World Is Not Enough* (1999): X-ray vision glasses, a hydroboat, a laser wristwatch, and a very high-tech BMW Z8 car. The car wins the Bond gadget award.
- *Die Another Day* (2002): a glass-shattering ring and perhaps the most unusual of all Bond cars, equipped with spikes for driving on ice and able to turn invisible with the touch of a button.

Today's Bond has superscience cars and superscience weapons. Yesterday's Bond had rowboats and guns. The technology is evolving, though perhaps not as quickly as we hope. Bond has evolved as a person, too: he's no longer quite the ladies' man that he was in the 1960s, and his M is now a woman, who won't put up with being slapped on the rump. The times are changing, and if Bond wants to remain working in this new century, he will, too.

2

Sending Secret Messages
Superspy Decoder Rings

My job is to kill you and deliver the Lektor.
How I do it is my business.
—Red Grant to 007, *From Russia with Love*

In the film *From Russia with Love* (1963), James Bond encounters the Russian Lektor message decoder. The Lektor is the size of a typewriter, weighs 22 pounds, and is housed in an ordinary brown case. According to the film, the Lektor has twenty-four "single" keys and sixteen "code" keys. A message goes into a slot on the Lektor, and then the decoded message appears on a paper roll on the side of the machine. Inside the Lektor are many perforated copper disks and small lights.

The Enigma Device

The Lektor was probably based on the German Enigma cipher device, in which agents fed data in one end and decoded messages came out the other end on a paper roll. This device was used in World War II, though its prototype was invented by the German scientist Arthur Scherbius in 1918. Scherbius's version was indeed based on a modified typewriter. Striking a key on this early Enigma turned a circular rotor that held twenty-six electrical contacts; these contacts corresponded to the letters of the alphabet. An agent would type his message in ordinary text on the early Enigma and the rotor would turn, and the resulting printed text was based on the electrical contacts hammered against the page.[1]

Later came the lights, one of which we see inside Bond's Lektor message decoder. In another version of the Enigma, the encoded characters were not hammered onto paper. Rather, each character was denoted by the lighting of one of twenty-six lamps on top of the Enigma machine. The position of the rotor as the agent struck the keys also determined which lamp was lit. So as the agent typed, the rotor position changed and the cipher (or encoded character) for the alphabet letters also changed. For example, if the agent typed four successive g's, the rotor positions would cause four different lamps to light. You can imagine the labor required to decode messages written in this fashion.

If you think this decoding mechanism is complicated, Scherbius's World War II–era Enigma was even more complex. That Enigma used four rotors. After the first rotor did a complete rotation, gears engaged the first rotor to advance to the second right from the middle of the alphabet on the second rotor. Then the second rotor engaged a third, again right in the middle of the alphabet. If you multiply $26 \times 26 \times 26$, you'll note that the Enigma could use a total of 17,576 characters to represent the characters in a plaintext message. The fourth rotor was not used to increase the total by another factor of 26. Instead, it was a reflector that sent an electrical signal based on the positions of the other three rotors. This electrical signal further complicated the cipher.

As long as both the transmitter and the receiver had all four rotors set up in an identical way, two Enigmas would successfully encode and then decode messages. The rotors were not interchangeable, because their internal wiring differed, so each Enigma had to use rotors fitted into the machine in a particular order. In addition, the starting position of each rotor had to be the same on both Enigmas.

The German Enigma excelled at being a superspy secret decoder ring—or we should say, four decoder rings were involved. The army and the Luftwaffe used three-rotor cipher machines rather than four. Basically, during World War II the Allies considered the German Enigma rotors to be invaluable gadgets. Luckily, the Allies were able to decode most messages that the Germans sent late in the war using Enigma devices.

Just how was the Enigma four-rotor code broken by the Allied forces? Basically, it involved banging out the problem and using process of elimination techniques. For example, some possible cipher combinations could be eliminated up front. Any character in the real plaintext message could be represented by any character other than itself—so that eliminated one possibility. Also, the Allied forces discovered that the plaintext character and its encoded counterpart had a reciprocal relationship; for example, if a plaintext character *b* was an encoded *W* at a specific rotor position, then the plaintext character *w* was an encoded *B* at the same rotor position.

Sometimes the Germans transmitted two versions of an encoded signal at the same time. British agents found that one of the two signals was always easy for them to decipher, while the other was always the more complex Enigma code. Using this technique, the British were able to decipher portions of Enigma code.

Beyond such simple matters, British agents ran endless combinations of plaintext and encoded characters through a Colossus computer, hoping that the computer would display a message that made sense. Colossus was located in a country house in Bletchley Park, Buckinghamshire. Among the people who created Colossus was Alan Turing, who later became famous for his role in early artificial intelligence. Turing's development of the Colossus machine and his analyses of the Enigma ciphers were instrumental in ending World War II. In 1945, Turing received the prestigious Order of the British Empire for his Colossus and Enigma decoding work during the war.

The Enigma didn't stop with the Germans and the British. Germany gave the Enigma to Japan, which used a modified version of the device. Americans called the Japanese Enigma Red. When the Japanese upgraded the Red boxes, the Americans called the new Enigmas Purple. American agents tapped into Japanese radio messages and cracked the Red codes. They did this by (1) brute force attempts to link Japanese forms of official address and phrases with often-used encoded terms and (2) getting lucky and intercepting plaintext messages along with the encoded ones.

U.S. Navy Intelligence intercepted a Japanese radio transmission on December 7, 1941, that instructed the Japanese ambassador to give a final warning to the United States by 1:00 P.M. (or 7:30 A.M. at Pearl Harbor). Intelligence agents decoded the full message at 8:00 A.M. and delivered the plaintext by 10:20 A.M. to Lieutenant Commander Alwin D. Kramer, who happened to be an expert on Japanese languages. Kramer determined that the message meant that the Japanese intended to attack the United States. With the Japanese a mere 300 miles from Pearl Harbor, General George Marshall received the decoded message in Washington, D.C. The general ordered that the message be encoded, then transmitted via the San Francisco RCA broadcasting system. American forces in Pearl Harbor received the message at 7:33 A.M. Simultaneously, the American radar systems in Pearl Harbor showed evidence of the Japanese attack.

This gadget, the Lektor aka Enigma, was a main plot device in *From Russia with Love*. As with many Bond devices in the 1960s, this one was clearly based on gadgets used during World War II. Bond was a bit behind the times. And as we'll soon see, the Bond cipher devices became more sophisticated in subsequent films, just as encryption technology advanced in the real world.

So, how did James Bond save the world with the Lektor device in *From Russia with Love*? Let's briefly review the plot. Bond goes to Istanbul to find the Russian encryption device. With the help of double agent Tatiana Romanova, Bond steals the Lektor and then has a heck of a time getting the device to London. Although he finally gets the Lektor/Enigma to London, we never learn in the film exactly how the device works and how the free world uses it to decode Russian messages.

The Clipper Chip

Moving forward in time to *GoldenEye* (1995), methods of encrypting and transmitting messages changed dramatically. For one thing, computers transformed from giant Colossus machines filling entire rooms to extremely powerful multiprocessing units linked via network systems. These newer computer configurations provided the power of Cray supercomputers, but then, of course, there were the Crays themselves.

By the time of *GoldenEye*, the Berlin Wall had collapsed, as had East European communism and the Soviet Union. Times had changed, and with them, so had Bond. He had become a secret agent for the new age. He no longer had to steal typewriter-sized decoder machines. Life was far more complicated for Mr. Bond.

At the start of the film, Bond opens a secure gate by pushing it (odd), and then he runs across the top of a dam. He shoots a grappling hook to the bottom of the dam by the Arkangel Chemical Weapons Facility in the Soviet Union. Using his rope and grappling hook, he jumps from the top of the dam over the cliff (a long jump, if you've seen the movie).

Bond's grappling hook doubles as a laser weapon that cuts through thick metal. Without goggles or any other protection, he uses the laser to break into the facility. With 006's assistance, he kills a scientist, then breaks into a main storage facility using a gadget that instantly opens a digital lock to the room. Steam is blowing everywhere in the main storage facility. Agents 007 and 006 stick a bomb timer on the wall; they're going to blow up the facility. Nobody's wearing gas masks or other protective gear: neither Bond nor the Soviet soldiers. This ends the opening scene of the movie—one of the best openings of all the Bond films, we think.

Without getting into details such as Xenia Onatopp, Bond's amazing camera and car (we talk about these items later in this book, but in case you're wondering, we don't devote an entire chapter to Xenia Onatopp), and real-time transmissions of video and audio to London from the Soviet Union and vice versa, we quickly realize while watching GoldenEye that this film is a technoscience gold mine compared to, say, *Dr. No* or *From Russia with Love*. Suffice it to say that if we have real-time video and audio feeds whipping around the globe, we've entered the modern era, and any encryption used in *GoldenEye* is going to far surpass the Enigma—oops, we mean Lektor—device in *From Russia with Love*.

Enter an imposter admiral, a crime syndicate, a high-tech helicopter, and GoldenEye, the Soviet weapon that beams electromagnetic pulses without affecting the high-tech helicopter. Now comes the encryption: we switch to the Space Weapons Control Center in Severnaya, Russia, which looks like Siberia (though Siberia is in western Russia and Severnaya is in central Russia). Boris Grishenko, a twenty-something Russian hacker, is hacking into the U.S. Department of Justice computers by decoding the U.S. Clipper Code. Does the Clipper Chip ring a bell?

The Clipper Chip was (and still is) a hardware encryption device created and used by the U.S. government right before *GoldenEye* appeared on screen. A crypotographic device, Clipper allowed secret agents to obtain an individual's keys to private transmissions. Clipper enabled agents to keep keys that deciphered communications between people on the Internet and through other forms of electronic communication.

The algorithm used to encrypt the transmissions was called Skipjack and was developed by the National Security Agency (NSA). Because the public objected to being spied on by government agents, Congress passed the Computer Security Act in 1987, limiting the NSA's jurisdiction in the surveillance of private citizens. Because the NSA classified its Skipjack algorithm as critical to national security, however, the agency continues to use Skipjack and the Clipper Chip. In fact, it was in 1993 that the White House officially announced the existence of the Clipper Chip. Then on February 4, 1994, Vice President Al Gore officially approved of government use of the Clipper Chip. Attorney General Janet Reno subsequently announced that the holders of the private keys would be the National Institute of Standards and Technology (NIST) and the U.S. Department of the Treasury. Reno also supplied a process by which law enforcement agents could obtain the private keys from the government and hence snoop on citizens. She supported the NSA claim that Skipjack and the Clipper Chip were needed for state and national security.[2]

So in *GoldenEye*, Grishenko quickly cracks the U.S. Clipper Code, which we assume means he disentangled the Skipjack algorithm. Frankly, it seems unreasonable that in 1995 a young fellow in Severnaya happens to uncover the Skipjack algorithm while attempting to hack into the U.S. Justice Department.

In response to the declassification of Skipjack by the Department of Defense in June 1998, the NIST published a description of Skipjack. In basic terms, to map 64-bit plaintext into a 64-bit ciphertext in 32 rounds, Skipjack uses an 80-bit key, $32 \times 4 = 128$ table lookup operations, and $32 \times 10 = 320$ XOR operations. Not so basic, is it? In fact, the published description was approximately twenty-two pages of math. It seems unlikely that Boris would be able to crack Skipjack within half an hour or less.[3]

Let's continue with our *GoldenEye* saga. Keep in mind that *GoldenEye* was a technology film for its time: it was rich in computer high-tech, and the inclusion of Skipjack was drawn directly from the newspapers.

Besides decoding Skipjack, Grishenko runs a program that seizes the phone line of anyone trying to trace him electronically. His program jams the other person's modem so he or she can't log off. And then Grishenko uses a traceback program to identify who is tracing him, which in this case means that the Federal Bureau of Investigation (FBI) screen appears on his computer screen. He now enters the FBI computer system.

How realistic is all this computer whizbang hacking? A traceback is a real method of finding out who's trying to hack into your computer. It evaluates the hops made from computer to computer throughout a network, looping its way back to the source. An example of a traceback is the Unix traceroute command, which uses something called the "time to live" field in an Internet protocol (IP) packet to obtain a "time exceeded" response from each node along the path leading back to the source computer. For example, a command such as:

```
# traceroute boris.siberia.org
```

might yield an output like this:

```
traceroute to boris.siberia.org (137.39.16.3), 60 hops max,
    40 byte packets
```

1 gateway (145.66.7.2) 10 ms 10 ms 10 ms 10 ms
2 yam.siberia.net (144.32.6.1) 250 ms 260 ms 250 ms 250 ms
3 flipper.ocean.net (135.78.0.3) 310 ms 310 ms 320 ms 310 ms
[many more computers along the way]
125 boris (137.39.16.3) 930 ms 840 ms 900 ms

In this example, the traceroute command tries to find the route that an IP packet follows to boris.siberia.org. The command sends probes over the network until the specified maximum number has occurred. Times are displayed that indicate how long it took for the probes to move from computer to computer along the way. This is a highly simplified view of traceroute, but you can see that Grishenko might actually be able to use this command in *GoldenEye* to detect that an FBI computer is hacking into his machine. It is a stretch, however, that he would be able to then hack into the FBI presto, just like that. But it gives us a great movie scene.

GoldenEye continues with murder and missiles in outer space. Electromagnetic radiation and enormous magnetic explosions give us action and adventure. Only one person supposedly survives the tremendous electromagnetic pulse explosion in Severnaya, and at this point in the film, it does not appear to be Grishenko. Rather, it is a female programmer, with no burn marks, no ripped clothing, and no bodily damage. The computers in Severnaya are fried. But this has nothing to do with encryption, so we'll leave *GoldenEye* for now. We'll return to it later in the book in our chapter on super-weapons.

Other Types of Encoding

One technique used often in Bond films is that of microencoding secret messages. For example, in *Live and Let Die*, Bond's hairbrush contains a Morse code transmitter. In *The Spy Who Loved Me*, Bond has a cigarette case that reads microfilm. He puts the microfilm into his cigarette lighter, slips the lighter into the case, and sees the microfilm images on a tiny screen that folds out of the cigarette case.

Microtransmitters and tiny microfilm readers are devices that have been commonly used for decades now. One aspect of microencoding messages that Bond does not use, however, is known in the real world as microdots.

This technique has been used since the Civil War era, and today's microdots are smaller than 1 millimeter in width. The cameras used to create microdots are often less than 1 inch long, and the pictures they produce are so small they can be concealed almost anywhere: on buttons, rulers, postcards, rings, stamps, and coins.

Tiny readers are used to view the images on microdots. These readers can also be concealed nearly anywhere, such as in pens and cigarettes. So Bond already has equipment that can be used to read microdot messages. Unless we've missed something, he hasn't used microdots yet, but maybe the use of this technology is coming.

More likely, though, is that Bond will emerge in his future films as an agent of today and tomorrow, meaning that his use of secret messages will include common technologies, such as the Internet and global satellite networks. Of course, he touched on the satellite networks a few times. In *A View to a Kill*, Bond encounters global satellite networks and world microchip domination in the form of Max Zorin, the head of high-tech Zorin Industries.

In today's world, the real surveillance comes in the form of data collection about people from every source imaginable: a credit card can reveal to a spy a person's name, address, employer, salary, spending habits, and bank accounts. Using that information, the spy can monitor and record all of a person's e-mail, telephone calls, and faxes. Web surfing reveals a person's interests, habits, and areas for potential exploitation and blackmail. Closed-circuit televisions are everywhere, recording our every move. In today's world, spies can remotely access our home computers and overhear our conversations simply by measuring the vibrations on our glass windows.

For more than a decade, secret messages have been encoded into digital images passed over the Internet. In fact, we use this technique to some extent in our 1999 techno-thriller novel *The Termination Node*. This is old technology now, but Bond hasn't used it yet.

The NSA has a system called Sombrero VI that holds a petabyte of data—that is, approximately 100 terabytes, eight times more data than is housed in the Library of Congress. The NSA is working on a bigger system designed to hold twenty times more data than the Sombrero VI. We don't understand why anyone would want to read ordinary drivel e-mail for spy-security reasons, but apparently this type of thing is actually happening. Perhaps Bond will someday use the resources of Sombrero VI, or its successors, to gather messages and unleash their secrets.

Or perhaps he'll use the resources of government cryptocomputing organizations. Because cryptography, or the encoding and decoding of secret messages, relies so much on supercomputing power, Bond's reliance on these government agencies is inevitable. For example, NSA computers are moving into speeds in the pentaflop area, that is, quadrillions of operations per second. It's expected that government computers soon will run at exaflop speeds, or quintillions of operations per second. Indeed, some people expect these computers to operate even faster, at, say, yottaflop speed (no, we're not making up *yottaflop*; it's a real word), equal to about a septillion (10^{24}) operations per second.[4]

Not only is the speed of information-gathering increasing at near-incomprehensible rates, but so is the miniaturization of computers. Will Bond have to ask for help from engineers who are running computer systems housed in gigantic buildings? Absolutely not. In fact, we expect Q Division to supply Bond with microscopic computers that can encode and decode messages within a second— or rather, within yottaflop speeds!

We've written in previous books about quantum and DNA computers. Recently, we've learned that government agencies are working on systems with seventy transistors that are embedded in a "cross-section of a human hair."[5] This leads us to believe that Bond will be able to wear his encryption/decryption devices on a strand of his own hair. Getting a haircut might be tricky in the near future, particularly for a secret agent.

3

Building a Bond Car

Flying, Underwater, and Missile-Launching Aston Martins

> *Need I remind you, 007, that you have a license to kill,*
> *not to break the traffic laws.*
> —Q TO 007, *GOLDENEYE*

Cars have become such an integral part of the James Bond legend that sometimes we forget that they weren't always so important. For example, in the first James Bond movie, *Dr. No* (1962), Bond drove a 1961 Sunbeam Alpine light blue convertible. It had no special gadgets nor did it play any role in the film. In an early scene in *From Russia with Love* (1963), Bond drove a Bentley Sports Tourer with a 4.5-liter engine, the favorite car of Bond's creator, Ian Fleming. Again, the car was just for show. Bond's cover when living in London was that of a wealthy man-about-town with a taste for the best things in life, and the cars he drove were merely part of his secret-agent disguise.

Luxury and Gadgets

That elite image changed in *Goldfinger*, which appeared in book form in 1959. In the novel, Bond's Bentley was retired from use and Q Division offered him a Jaguar 3.4 or an Aston Martin DB3. When Bond selected the new silver metallic Aston Martin, the history of modern spy novels changed forever.

It was *Goldfinger* that popularized in books, and later in movies, the notion that luxury cars, modified with special devices, were essential to spies. The idea had been used earlier by the popular English author Dornford Yates in his Richard Chandos thrillers; those novels may have served as Fleming's inspiration. In the United States, Batman used a special car equipped with innumerable gadgets, but it's unlikely that Fleming thought of Batman comic books when creating Bond.

In *Goldfinger*, the Aston Martin selected by Bond had been rebuilt by Special Branch to include a number of extras, including reinforced steel bumpers, trick lights, and a handgun hidden in a secret compartment. In the film, the modifications were more complex, and the car was a DB5 instead of the already dated DB3. The movie-version Aston Martin contained two machine guns hidden behind the parking lights, and it fired smoke canisters out of the exhaust system. A sliding steel plate in the rear of the car provided protection from enemy bullets, while tire cutters extended from the car hubcaps. Along with gadgets that sprayed oil and deposited nails on the road, the car also came equipped with an ejection seat for the side passenger. The Aston Martin DB5 proved so popular with audiences that even though it crashed into a building in *Goldfinger*, it reappeared in the next Bond film, *Thunderball*.

The only problem with Bond's special car was that it was both impractical and impossible. Impractical because any steel sliding plate in the rear of the car needed to be incredibly thin. Otherwise, the weight of the plate would act as an anchor, slowing the car down to a crawl. Plus, the weight of the panel would make it impossible for it to slide up and down with the ease displayed in the movie. A steel plate thin enough to work right, unfortunately, wouldn't be able to stop a bullet, rendering it useless as a shield.

An ejection seat designed to send a passenger into the air would be equally impracticable; its weight and mechanisms would be better suited for a tank than for an Aston Martin. It was such an outrageous gimmick and so ridiculous looking that Bond designers left it out of all future versions of Bond cars.

Not nearly as visually exciting but much more practical in the movie was the tracking system used by Bond to follow Goldfinger's car from England to the Alps. A small homing device planted on the underside of the villain's auto transmitted a signal that Bond tracked using a dashboard map on the Aston Martin. Supposedly, CIA chief Allen Dulles expressed interest in whether the device existed and how it worked. While homing devices weren't new, ones that emitted a signal that could be tracked for hundreds of miles were well beyond 1960s technology. But, unlike the outrageous gadgets included in the car, such as machine guns and oil-slick spreaders, high-tech tracking devices developed into one of the most useful devices ever introduced to the general public via a James Bond movie.

The history of the Global Positioning System began when the Soviets launched the first space satellite, *Sputnik*, in October 1957. Researchers at the Massachusetts Institute of Technology soon realized that they were able to calculate the orbit of the satellite by tracking its radio signal. This discovery led to the notion that a person's position could be determined by using radio signals beamed from a satellite.

Taking that notion one step further, the U.S. Navy in 1967, only a few years after *Goldfinger* appeared in theaters, developed what was known as the Transit System for submarines armed with Polaris nuclear missiles. This system of six satellites circled the earth in polar orbits. Using the satellites for positioning, the submarines could locate their positions anywhere in the world in fifteen minutes.

In 1973, the U.S. Department of Defense decided to expand the Transit System into a global network of satellites that would provide the location for any military vehicle anywhere in the world. The system cost $12 billion to put into operation. The satellite network was originally called the Navstar Global Positioning System, but its name was later changed to the Global Positioning System (GPS). A network of twenty-four satellites, the GPS was officially started with the launch of its first satellite in 1978. The system was completed with the orbiting of the twenty-fourth satellite in 1994. In 1984, President Ronald Reagan announced that the system would be available for civilian as well as military use. As the satellites do wear out over time, the fiftieth GPS satellite was launched in March 2004.

The GPS works by triangulation. By very accurately measuring our distance from three GPS satellites, we can calculate our position anywhere on Earth. Each of the three measurements gives us a sphere on whose circumference we must be located. Intersecting the circumferences of the three spheres determined by the three satellites gives us two common points. One of these two points always turns out to be impossible, thus determining that the other point is our precise location. A fourth satellite is usually included in the calculations to determine the exact time the measurement is taken. All GPS satellites carry atomic clocks for precise time.

At first, the GPS required sophisticated instruments to locate military equipment or buildings, but like everything else in today's world, technology soon shrank the size of receivers. In 1998, GPS receivers were small enough to be included in a wireless phone. At present, most PDAs (personal digital assistants), portable phones, and laptop computers are equipped with GPS chips. In a few years, such chips will not only provide navigation information for most automobiles but will also make tracking stolen cars possible with the touch of a switch. Amazing in 1964, the GPS is another facet of modern technology predicted in casual fashion by the James Bond films.

The DB5 car was so popular in *Goldfinger* that Aston Martin was deluged with requests to put the vehicle on display. Realizing the tremendous publicity potential, the company built two replicas of the car for promotional purposes. The cars were equipped with some unusual extras. One had a telephone that was installed in the driver's side door. It also had luxurious antelope leather interior trim as well as a special reserve gas tank. All three of the special cars, as well as two more that had been used in filming *Goldfinger*, were put on display at auto shows and charity events. They served as the best publicity that Aston Martin ever had and made it clear to the automobile industry the advantage of being featured in a James Bond movie.

The follow-up film to *Goldfinger* was *Thunderball*, and the Aston Martin DB5 showed up at the beginning of the new movie. Bond uses several of the car's gadgets to escape a group of gangsters during the pretitle sequence that takes place outside the Château d'Anet palace near Paris. Since *Thunderball* took place primarily in the Caribbean and underwater, there weren't any other opportunities for Bond to use an automobile. The Aston Martin DB5 also made brief appearances in *GoldenEye* (1995) and *Tomorrow Never Dies* (1997).

In *You Only Live Twice*, Bond never gets a chance to drive a car. However, a female agent of the Japanese secret service named Aki manages to help Bond by driving a Toyota 2000 GT, propelled by an inline six-cylinder engine with two camshafts that enable the car to reach a top speed of approximately 140 miles per hour. Toyota produced 350 of these cars, with two of them transformed into convertibles for the movie. Electronic gadgets made the car unique. The car had a closed-circuit TV system with cameras behind the license plates to record everything in front of or behind the car. Also included was a cordless telephone, a voice-activated cassette player, a video recorder in the glove compartment, and a miniature color TV, typical items for today's luxury cars, but something very special for the 1967 world of James Bond.

While car chases and expensive cars played important roles in the next three Bond movies, none of them featured much in the way of unconventional equipment or unique devices, unless we include the experimental silver-white moonbuggy that Bond used to escape Ernst Stavro Blofeld's agents in *Diamonds Are Forever*. It wasn't until 1974's *The Man with the Golden Gun* that a Bond car used modern science in unique fashion. The only problem with that car was that it belonged not to James Bond but to his enemy, Scaramanga.

Flying Cars

In *The Man with the Golden Gun*, James Bond chases the villain Scaramanga through the streets of Bangkok to an airstrip, only to have the notorious assassin escape capture by converting his car into an airplane and flying away. Bond is left empty-handed, while his nemesis wings his way to his hideaway on an island in the China

Sea. Needless to say, Bond soon finds his own aircraft and flies off for his final duel with Scaramanga, leaving the audience and us to wonder if it's possible for a car to change so quickly into an airplane, and, if it is, where has this car been hiding during rush hour?

In the film, the flying car was an AMC Matador Coupe. With a flight tail unit added on, it was approximately 30 feet long with a wingspan of approximately 40 feet. In the movie, it supposedly flew from Bangkok to an island in the China Sea, but in real life, the vehicle used for filming was capable of flying only 1,500 feet. The problem, however, wasn't with the concept; it was just that Scaramanga had picked the wrong model.

The idea of flying cars wasn't new to the film. Many science-fiction novels from the late nineteenth and early twentieth century imagined a future where people no longer drove along highways but instead flew everywhere. Giant floating cities were featured regularly in European and American sci-fi stories. Clifford Simak's famous series of novelettes, *City*, published in the early 1940s, detailed how America changed from urban to rural living due to home airplanes that made commuting long distances relatively quick. Recent science-fiction films such as *The Fifth Element* and *Star Wars: Episode II—Attack of the Clones* show a future where gigantic cities feature multiple levels of continuous air transport.

The invention of the airplane in 1903 by the Wright brothers, along with America's fascination with automobiles, prompted inventors to try combining the two ideas into one machine. According to the U.S. Patent and Trademark Office, nearly eighty patents for flying cars were registered in the past century. Unfortunately, most of the ideas were impractical and didn't work. The few that did show some small signs of success never proved practical enough to attract major financing.[1]

To fully understand how to make a car fly, we need to understand the basics of aerodynamics, which means we must define terms related to the subject.

Thrust is the force generated by a plane or a bird to move forward. Airplanes in the twenty-first century normally create thrust by using jet engines. Helicopters and autogyros (remember this for later) create thrust by spinning rotor blades above the craft. Birds create thrust by flapping their wings. The aerodynamic force overcome by thrust is known as *drag*.

The third force acting on a flying object is its *weight*. One fact worth remembering is that everything on Earth, including air, has weight. Obviously, the weight of a plane or a helicopter is what keeps it on the ground. The aerodynamic force that raises and holds the airplane or helicopter in the air is known as *lift*. When lift is greater than weight, a plane flies. When lift decreases, the plane descends. Both lift and drag can exist only in a moving fluid, and air, in this case, is considered to be a fluid. Thus, neither airplanes nor helicopters can work in outer space, where there is no fluid.

Of these four terms, the first three are easy to understand. A motor powers an airplane, developing thrust. Air resistance acts as drag. The weight of an object is its mass affected by the power of gravity. Only lift is a mystery. How is lift created?

There are two common explanations for lift found in most textbooks. The first is Bernoulli's principle, while the other is the transfer of momentum principle. Each of these standard explanations is partially correct, but each is also partially incorrect. Even though lift is not so easily explained, we'll do our best.

Bernoulli's principle is often called the "longer path" explanation. It looks at the top and bottom surfaces of an airplane wing and deals with a stream of air particles traveling toward the front of the wing. A wing is constructed so that the top is raised and curved. Therefore, when the air particles split at the wing tip, the ones traveling over the top have a greater distance to travel to the back. If they are going to take the same amount of time to make it to the back of the wing as the particles going beneath, the particles going above the wing must travel faster. Bernoulli's principle, one of the fundamental rules of fluid dynamics, says that as the speed of a fluid increases, the pressure it exerts decreases. Thus, Bernoulli's principle implies that the pressure on the top of the wing must be less than the pressure on the bottom of the wing. Since the air pressure beneath the wing is greater than the air pressure above the wing, the wing (and with it, the airplane) rises.

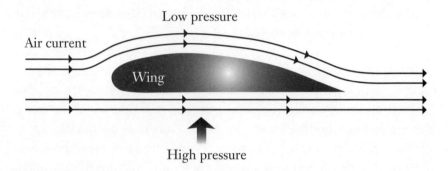

Unfortunately, there are several major problems associated with applying Bernoulli's principle to wings. For one, there's no valid reason why the air particles that go over and under the wing need to meet at the back of the wing at the same time. Another problem is that not all wings have a curve on top and a flat surface on the bottom. Some wings are curved both above and below. More troublesome is that sometimes planes fly upside down. If Bernoulli's principle is true all the time, the higher air pressure would be pressing down on the top of the wing and would drive the plane straight into the ground.

Still, the longer path explanation isn't entirely wrong. The air on the top part of the wing does flow faster than on the bottom of the wing. So there is some truth to the theory.

The transfer of momentum principle is based on Isaac Newton's third law, that for every action there is an equal and opposite reaction. Newton imagined air molecules acting like bullets and striking the bottom surface of the wing. These particles would add the force of momentum to the wing and slowly move it up into the air.

The problem with Newton's idea is that air acts as a fluid, not as a stream of molecules. Also, his theory never takes into account the top surface of the wing, so his calculations are not very accurate. At very high speeds (five times that of sound), however, air molecules behave more like bullets, so Newton's ideas aren't entirely worthless.

If we take the best of both theories, we finally come up with a clear and concise explanation for lift. Lift is a force on a wing completely immersed in a moving fluid (air). It acts on the wing in a direction perpendicular to the flow of the air. The force is created by differences in pressure that occur because of differences in the speed of the air all around the wing (both top and bottom). The result of this force is divided into lift (raising the wing) and drag (slowing it down). When the flow of air past a wing is increased, the pressure differences between the top and the bottom of the wing become greater, and lift increases. Lift can also be changed by varying the angle of the wing.

Putting all these factors together, we come up with the standard equation used for calculating lift:

$$L = \tfrac{1}{2} \times C_1 \times R \times V^2 \times A$$

where

L = lift
C_1 = the lift coefficient
R = air density
V = air velocity
A = wing area

The lift coefficient is a number entirely dependent on the angle of the wing, while the air density is totally dependent on the height of the plane above sea level.

Examining the equation, we immediately deduce that lift depends on two variables: air velocity and wing area. Thus, the faster a plane is moving and the bigger the wings, the more lift it generates, leading us to conclude that a flying car needs a large motor and big wings.

The first airplane-car was invented by Glenn Curtiss in 1917. Curtiss built his Autoplane from aluminum. The machine had three wings with a total wingspan of 40 feet. A large four-bladed propeller run by the car's engine was placed at the rear of the Autoplane. The vehicle never actually flew, but it did manage to get off the ground a few times.[2]

A much more ambitious effort was the Airphibian, the invention of Robert Fulton, which made its first appearance in 1946. Fulton abandoned the usual idea for car-plane hybrids, where a car was constructed to fly, and instead built a plane that could ride on the highway. The Airphibian featured wings and a tail section that could be removed from the plane, with the propeller kept safe inside the plane's fuselage. According to the designer, it took less than five minutes to transform the airplane into a car. The Airphibian could fly at 120 miles per hour and drive on the road at 50 mph.[3] Despite his invention's success, Fulton was never able to find a company willing to finance his airplane. One of the major problems with the Airphibian was that in car mode it looked like a small airplane with its wings and tail knocked off. The vehicle might have been practical, but it looked ridiculous.

The most successful combination of car and plane was Moulton Taylor's Aerocar. Taylor had met Fulton years before and was an admirer of the Airphibian. Learning from Fulton's mistakes, Taylor designed his vehicle to ride, then fly, then ride again without major changes. The car's weight was kept down by the use of a fiberglass body. The entire machine was 26 feet long, with a wingspan of 34 feet and a wing area of 190 square feet; it weighed only 2,100 pounds. The plane had a cruising speed of 135 mph in the air and could reach a ceiling of 12,000 feet. On the ground, it had a cruising speed of 60 mph. In the air, the Aerocar functioned perfectly as an airplane. On the ground, it took fifteen minutes to remove the wings and tail, which were left at the airport while the car went cruising. The Aerocar attracted interest from several big automakers, including Ford, but it never went into production.[4]

Without question, the vehicle used by Scaramanga in *The Man with the Golden Gun* was an Aerocar. Taylor and Fulton would have been proud.

Diving Cars

Moving from air and land, we naturally turn to air and sea. In the tenth Bond film, *The Spy Who Loved Me*, Bond fights a billionaire who dreams of ruling the world's oceans while destroying all humanity on land. With so much of the film taking place underwater, it was only natural for Bond to drive a car equipped to travel both on land and under the waves.

Knowing the huge impact a Bond movie could have on new car sales, the manager of Lotus cars parked the latest model, the Esprit, designed by Giorgio Giugiaro, near the front gate of Pinewood Studios in London in 1975. Cubby Broccoli, the man behind the Bond films, was suitably impressed, and the Lotus Esprit became the Bond car for *The Spy Who Loved Me*. Until *Die Another Day*, the Lotus was the most fully equipped and outrageous car ever driven by James Bond.

Among its essentials, the car had four rear sprayers behind the license plate that fired cement at pursuers. When the car entered the water, it converted into a fully manageable two-person submarine. While underwater, it was capable of discharging floating mines and underwater color smoke screens. Like a submarine, it fired torpedoes, and when threatened by a helicopter hovering over its position, the car discharged a surface-to-air missile.

Unfortunately, while the Lotus looked impressive converting into a submarine (the wheels folded up inside, and fins and a periscope emerged), the car was deemed impossible at the time, especially considering the speed and ease with which it moved through water and its sleek, almost manta ray look. No one suspected that less than twenty years later a portable submersible vehicle resembling another sea creature, the dolphin, would make its real-life debut in California.

Tarco Research was founded in 1977, when the film *The Spy Who Loved Me* was a huge James Bond hit. Started by Thomas A. "Doc" Rowe, the company's mission was to design and develop advanced concept vehicles, especially those for underwater use. These new over-and-underwater boats were based on Rowe's concept of Variable Altitude Submersible Hydrofoil technology, which he first envisioned in 1974. After many years of hard work and experimentation, Tarco unveiled the first Bionic Dolphin at Whiskey Town Lake, California, on September 7, 1992. According to Rowe, the boat was the first lighter-than-water vehicle to successfully move underwater.[5] *Discover* magazine called the Bionic

Dolphin's run "the aqueous parallel to the Wright Brothers' achievement: underwater flight."[6]

As each Bionic Dolphin is individually designed to the owner's wishes (the boat was featured on the cover of the fall 1996 Hammacher Schlemmer catalog, selling for only $139,000), it could easily have been bought and modified by Q Division for Bond's use, especially if they combined the features of the Bionic Dolphin with those of the Rinspeed Splash or the Bond Aquada.

The Rinspeed Splash is the latest sports vehicle designed by Frank M. Rinderknecht, the founder of Rinspeed Technology. A complex hydrofoil system allows a flashy sports car to transform into a speedboat that flies almost a foot above the water. The transformation from automobile to boat is done by a complex computer-controlled hydraulic system. The custom-designed transfer case sends power to the car's rear wheels or propeller or both, depending on how far into the water the Splash has gone. At a water depth of about 1 meter, the car transforms entirely into a boat. A complex system of hydrofoils integrated into the design of the car rotate down and under the chassis. Then the Z-drive and propeller take over. Even at low speeds, the hydrofoils start lifting the car-boat out of the water. At full power, the Splash travels at 52 mph.[7] At present, the Splash is not available for sale. But the Bond Aquada is.

Designed by Gibbs Technologies, the Bond Aquada sports car and speedboat combination has a cruising speed of 100 mph on land. When it drives into the water, the wheels retract into the wheel arch, power jets turn on, and the car converts to a boat. On the water, it cruises at 30 mph.

The car has no doors, a design feature to prevent leaks. According to Alan Gibbs, the chairman of Gibbs Technologies, the Aquada is the product of a seven-year development program and sixty newly patented technologies. The car-boat is priced at $235,000.[8]

The Aquada was prominently featured in the news on June 14, 2004, when it set a speed record on water. The millionaire adventurer Richard Branson drove across the English Channel in the amphibious sports car. He crossed the twenty-two-mile-wide channel from England to Calais, France, in one hour, forty minutes, and six seconds.

Together, a Bionic Dolphin and a Bond Aquada would cost less than $400,000, a small price to pay for saving the world from a madman who intends to wipe out most of the human race.

Bond Cars for the New Century

The Lotus Esprit Turbo played a small role in the next Bond film, *For Your Eyes Only*. An enemy agent finds Bond's car and tries to tamper with the lock. The car has the best possible antitheft device: it self-destructs, leaving Bond to rely on other people's cars for the rest of the film.

After that film, specialized cars hardly appeared in the next few Bond films. It wasn't until Timothy Dalton took over as Bond in *The Living Daylights* that well-equipped vehicles returned. In Dalton's first Bond film, he drove an Aston Martin V8 Vantage. The car had laser guns in the wheels, outriggers for skiing on ice, self-inflating spiked tires, and a jet propulsion system. All these devices proved useful in a drive across a frozen lake while being pursued by the Austrian police.

In *GoldenEye*, the Bond producers decided to use, for the first time in the series, a car not manufactured in England. Instead, they chose as the first ever official Bond car the BMW Z3 Roadster. A major marketing agreement called for BMW to promote the film in their worldwide campaign for the new car from November 1995 through March 1996. No money was involved in the deal, which became a major cross-promotion for the film and the car.

On November 13, 1995, *GoldenEye* premiered at Radio City Music Hall in New York City. The Z3 Roadster was displayed for the first time to TV and newspaper reporters in Central Park by Desmond Llewelyn (Q) and Pierce Brosnan (James Bond). That night, twenty Z3 cars were parked near Radio City for examination by the rich and famous attending the premiere. The car was a huge success, with buyers including Alec Baldwin, Madonna, and Steven Spielberg.

In *GoldenEye*, the Z3 was mostly for show and had little to do with the action. In the next Bond film, *Tomorrow Never Dies*, Q arranged for Bond to drive a BMW 750, and this time the agent put the car through its paces. As before, it was fully loaded with weapons, including machine guns and rocket launchers, but what made the new car so special was a remote control hidden in Bond's portable telephone. Bond used the remote control when trapped in the backseat of the car while being chased by enemy agents. After exiting the car, Bond, working the remote control, led the villains on a high-speed chase through a multilevel parking garage that ended with his car crashing into an Avis rental agency across the street.

The engineers working for Q didn't rely on scientific break-throughs to construct Bond's remote control unit for the BMW. The technology had been on the market for years, and it wasn't just for spies.

The AutoCommand Professional Series Remote Car Starter from Gizmocity.com enables a driver to start his car from up to 500 feet away. The keychain transmitter not only starts the vehicle by remote control but also locks and unlocks the car doors, activates the car alarm, opens the trunk, and turns on the heat or air condi-tioning.[9]

Equally useful are the electronic controls available for full-feature remote-controlled model racing cars that can be purchased from hundreds of Internet sites. Controlling a real car with a remote control is child's play, figuratively and literally, for amateur racers and secret agents.

More to the point, Q's mechanics could always watch the Discovery Channel's car show *Monster Garage* for ideas. On the December 1, 2003, episode, the builders converted an out-of-commission 2003 Chevrolet Monte Carlo figure-8 race car into a life-sized remote-control racer in less than a week, and for only a few thousand dollars.

While building a remote control for a BMW might not be too difficult, turning an Aston Martin invisible takes some thought and a lot of computer hardware, as demonstrated in 2002's Bond hit, *Die Another Day*.

After driving BMWs in three films, Bond returns to a British car, an Aston Martin V12 Vanquish, in *Die Another Day*. In keeping with tradition as established for all the Pierce Brosnan Bond adventures, the Vanquish is equipped with all the latest spy gadgets, including four grille-mounted rockets, a pair of machine guns, and two motion-detector guns. The spy car also has special spiked tires for driving on ice, as a good part of the film takes place in frozen regions. But the most amazing feature of the car is that with the flick of a switch, it can be turned invisible. Equally impressive is that the invisibility is not magic but actual state-of-the-art technology.

Optical camouflage is a new kind of camouflage made possible through the use of microcomputers. It can best be called active camouflage. The basic concept is simple: If you photograph a background and then project that background onto a screen, the object behind the screen appears to be transparent.[10] The Japanese scientist who perfected this unique form of camouflage, Dr. Sekiguchi S. Tachi, calls his invention X'tal Vision. He demonstrated X'tal Vision using what he calls a "transparent cloak." When wearing the cloak, which resembles an ordinary raincoat, a person appears to be totally transparent. Cars, trees, and people who are behind the subject can be seen clearly, thus creating the illusion that the wearer is invisible.

In *Die Another Day*, the principle of the transparent cloak is used for an entire automobile. Thousands of tiny digital video cameras are embedded in the framework of the car. These cameras take pictures of the nearby scenery and transmit the images to projectors on the opposite side of the car. The projectors display the images as photographed by the cameras. The thousands of small images combine to form one large digital video of what is just beyond the car. Thus, the vehicle is hidden behind a screen of scenery. The image quality, of course, depends a great deal on the background scenery. If used on a busy street in Manhattan, the invisible car would most likely be a disaster, since it would project people behind the car moving in front of it. In the Bond film, the main action scenes involving the invisible car take place in an icy wilderness, so the car blends with the background.

In the end, the invisible car provides some tricks and kicks in *Die Another Day*, but it's really just another interesting gimmick in the movie. Invisible cars are interesting developments, but for the reasonably near future, there seems to be no chance that we'll be seeing them on city streets. This is an idea whose time has yet to come. As with most of the specialized James Bond cars, it has a unique standing in the world of espionage, but as with every machine, it's only as good as its driver.

4

Arming Yourself

(and Other Explosive Ideas)

If you carry a 00 number, it's a license to kill, not be killed.
—M TO JAMES BOND, *DR. NO*

*. . . the Walther PPK, 7.65 mm with a delivery like a
brick through a plate glass window.*
—BOOTHROYD, THE GUN MASTER, TO JAMES BOND, *DR. NO*

It is in James Bond's first movie adventure, *Dr. No*, that he receives his first Walther PPK gun. And he continues to use the Walther until *Tomorrow Never Dies*, when he switches to the Walther P99. Actually, Bond begins in *Dr. No* with a Beretta, which he claims to have been using for ten years. M lectures him about his Beretta problems, such as the mission during which the gun jammed, putting Bond in the hospital for six months. Hence M declares that "if you carry a 00 number, it's a license to kill, not be killed." And hence the gun master supplies the Walther PPK, extolling its virtues.

Bond's Guns

The Walther PPK pistol is designed for German police (by the German Walther Company, which was founded in 1886) and is very popular with secret agents around the world. It is effective, yet small, compact, and easy to hide.

In fact, the Walther PPK is only 1.08 inches wide, with a barrel length of 3 inches and an overall length of 6 inches. It weighs 21 ounces and has either a stainless steel or (more traditionally) a blue finish. It has double action, and is available as a .380 ACP or .32 ACP caliber pistol.

The Polizei Pistole Kriminal (PPK) includes a signal pin that shows whether the gun is loaded. If the chamber is loaded, the signal pin protrudes from the back of the slide. If the chamber isn't loaded, the signal pin cannot be seen, as it remains inside the slide.

As mentioned in *Dr. No*, the rounds are 7.65 mm—we'll explain all these gun-related terms, we promise!—though permutations of the Walther allow for different rounds. While the PPK isn't available in the United States, Americans can obtain the PPK/S. Around the world—in England, Germany, France, and Israel, to name a few countries—the PPK is commonly used by intelligence agencies.

Caliber is the measure of the internal barrel width of a gun. A .45 caliber gun has a .45-inch barrel width. While the caliber usually does measure the internal barrel width, some deviations exist. For example, a .38 Special revolver actually has a .36 caliber barrel and shoots the same bullet as a .357 Magnum revolver. A 7.65 mm round means that the cartridge has a diameter of 7.65 mm.

Bullets come in different calibers. The caliber in this case indicates the diameter of the bullet. Some common bullet calibers are: .22, .25, .357, .38, .44, and .45.

Revolvers, which have rotating barrels to hold the bullets, are generally .38 Special or .44 Magnum caliber. Bond's Walther PPK, however, does not have a rotating barrel to hold bullets. Rather, the Walther is a semiautomatic pistol. When we say that the Walther has .380 ACP or .32 ACP, we mean that it is a .380 or .32 Automatic Colt Pistol. The ACP part of the caliber definition tells us that the Walther is a semiautomatic pistol with a particular shape and length of bullet chamber. The larger the barrel, the larger the bullet, the larger the impact.

The words *magnum* and *super* in caliber designations indicate that the cartridge has more projectile energy than cartridges without these designations. The .357 Magnum is a caliber for a cartridge that is very similar to the .38 Special except that its case is longer and contains more gunpowder, and therefore produces greater projectile energy than the .38 Special. The .38 Special is the caliber for the revolver that was long the standard weapon for police and for civilian self-protection. It and the .357 Magnum are still much used in this manner.

A .357 Magnum revolver can shoot .38 Special cartridges in addition to the "mag" cartridges. This is often done for target practice, because the smaller cartridges are cheaper and don't kick as hard. The .357 Magnum is also used for hunting by people who like to give the animal more of a chance than it would get from a hunter with a rifle (the handgun hunter has to get closer, because the handgun is not as accurate as a rifle).

A cartridge with a very slightly smaller outside diameter than the .38 Special is the .380 Auto for automatic and semiautomatic handguns. The .380 cartridge case is shorter than that of the .38 Special and holds less powder, so the cartridge does not produce as much energy as the .38 Special.

Some of the inch calibers mentioned earlier generally refer only to cartridges for automatic or semiautomatic handguns. For example, the .25 and .380 are common calibers for handguns but not for revolvers. Other calibers are common only for revolvers (like .38 Special and .44 Magnum). For some, however, there are cartridges for both types of guns. An example is .45 Colt for revolver and .45 ACP for automatic and semiautomatic. For some numerical designations there may be several different calibers differentiated by the appended abbreviation.

It makes sense for James Bond to carry the type of gun used by spies across the globe. A .25 caliber pistol wouldn't defend him as well as his .380 (or .32) Walther, and something as macho-massive as a .44 Magnum revolver—used by Clint Eastwood in the Dirty Harry movies—weighs as much as a six-pack of beer, has a 6-inch barrel, and is not particularly easy to hide in a vest (or tuxedo jacket) pocket.

A single-action gun means that the shooter cocks the gun and squeezes the trigger. The original meaning of a double-action revolver was one that could be fired either by cocking the gun and squeezing the trigger, or by using a long trigger pull, thus raising the hammer and dropping it to fire the bullet. In today's world, a double-action gun can be one that is shot using a long trigger pull.

It's slightly different with a semiautomatic pistol. When Bond fires his Walther PPK, he needs a long, heavy pull on the trigger, which raises the hammer fully, snapping the hammer to the fire pin, which then hits the primer. The primer ignites the powder charge, hence propelling the bullet from the chamber.

After firing the PPK once, it becomes much easier for Bond— not that he ever needs help firing his gun the first time. After the first shot, the burning powder pushes the slide toward the back of the gun. The slide carries the empty cartridge case and ejects it. In addition, as the slide moves, it cocks the hammer again. Then as the slide shifts forward into its original position, it inserts a new round from the magazine into the chamber. Now Bond's gun is automatically cocked and ready to fire another bullet. He squeezes the trigger, this time much more lightly.

The rounds of bullets are in a detachable magazine, which we just mentioned. *Magazine* is a strange word for this container. It really just looks like a metal holder for bullets, and it slides into the gun where Bond is gripping it. The hammer is on the back top of the gun, so when Bond grips the PPK, his thumb can be on the hammer and his forefinger on the trigger. The ejection port, where the rounds are expelled during Bond's shooting sprees, is on the top of the gun above the trigger.

Of course, in the real world spies use guns other than Walther PPKs. For example, the 7.65 mm CZ27 semiautomatic gun with silencer was made by the Czechs during World War II and was used after the war by German agents. Other silencer guns from World War II, such as the Special Operation Executive's Welrod and the Office of Strategic Service's Hi-Standard Pistol, are still being used by agents today.

By the way, have you ever wondered why Q Division lets Bond use an ordinary gun rather than issuing him a custom pistol that includes X-ray vision filming, fingerprint analysis, genetics decomposition, and satellite transmission capabilities? The reason is simple: intelligence agencies tend to issue standard types of guns so the weapons can't be traced back to them.

Suppose a secret agent—not James Bond, of course—assassinates the leader of another country. The agent drops his weapon and flees the scene of the crime. His weapon happens to be a flame-throwing, blood-analyzing, parachute-enabling, micro-encrypting, data-transmitting, voice-recognizing .22 caliber (bad choice in caliber for an assassin) pistol that looks like an ice pick. Aha, the authorities exclaim as they pick up the ice-pick weapon lying on the ground by the murdered leader; only James Bond (or Agent X from Country Y) carries such a weird weapon!

But although a special-issue weapon may seem to be the same as a standard-issue weapon, the former has a slightly better trigger or provides slightly better quality during aim.

Exploding Wristwatches

Let's move beyond the guns. There's far more to James Bond and explosives than firearms. For example, there are wristwatches that explode and wristwatches that function as detonators. In fact, there are wristwatches with Geiger counters and ticker-tape readouts. Here's a brief recap:

- *From Russia with Love*: Red Grant, the deadly SPECTRE assassin, wears a watch that contains a piano-wire garrote cord. In 007's battle with Grant, Bond uses the wire to strangle his attacker. The watch is seen, though not used, in two later Bond movies, *A View to a Kill* and *Licence to Kill*.
- *Live and Let Die*: Bond's Rolex wristwatch includes an electromagnet and a circular saw. From far away, the electromagnet deflects the paths of bullets. It also attracts metal to the watch, and Bond uses this feature to magnetize M's spoon. When Bond spins the face of the watch, it turns into a circular saw, enabling him to cut through ropes and escape capture.

  In addition, Bond has a digital watch that displays the time when the doorbell rings. Not very interesting by today's standards, but pretty cool at the time.

- *The Spy Who Loved Me*: Bond's wristwatch has a ticker-tape readout. While Bond is in Austria, the watch prints a ticker-tape message with mission instructions from headquarters.
- *Moonraker*: Bond's Seiko watch not only has a detonator but also includes a circular pad of explosives beneath the watch face. Bond uses a piton and a piece of wire to activate the explosives, then presses a button on the watch to detonate the device.

  In addition, Bond has a dart-gun wristwatch—sort of. Basically, a dart gun is strapped onto his wrist so that it looks like a watch, and then the darts are activated from the current in Bond's nervous system. This is an extremely dangerous wristband, given that it comes with ten darts: five that can pierce armor and five that are coated with cyanide. If Bond is feeling jumpy, anyone nearby could get killed within a minute.

  It's a good thing that Bond has the dart-gun band, however. When he's in Dr. Holly Goodhead's centrifuge room and Chang is trying to kill him using intense *g*-forces, Bond uses the dart gun to short-circuit the *g*-force control system. Later, as Hugo Drax is escaping, he finds a gun and threatens Bond with it, but thanks to Q Division, Bond has his handy wristwatch and shoots a dart at Drax.
- *For Your Eyes Only*: Bond's wristwatch is a communications device but does not provide him with explosives, poison darts, electromagnets, or other weapons. Digital messages scroll across the face of the watch, and Bond uses the watch as a cell phone.
- *Octopussy*: Bond's digital wristwatch includes a tracker for a homing device that's in a fake Fabergé egg. Of course, the egg is wired for sound, and Bond picks up the audio using his Mont Blanc fountain pen.

  Bond also has a liquid crystal television watch that he uses in Q's lab to ogle a girl. Not as important as ogling the girl,

Bond also uses the watch along with cameras installed in Kamal Khan's palace to do his job. In this second case, the cameras record Khan and Octopussy leaving the palace through a secret escape route and transmit the video to Bond's watch.

- *A View to a Kill*: Bond's wristwatch includes a garrote cord.
- *GoldenEye*: Bond's Omega Seamaster wristwatch has a laser and detonates explosives. The laser is powerful enough to cut through the floor of Alec Trevelyan's missile train. And, as in *Tomorrow Never Dies*, the wristwatch has a special button that detonates explosives. When Bond attaches remote-controlled explosive mines to barrels containing flammable liquid, Trevelyan deactivates the explosives by unknowingly pressing the button on Bond's watch.
- *Tomorrow Never Dies*: Bond's Omega Seamaster wristwatch has a special button that detonates explosives. While on Elliot Carver's stealth boat, Bond activates the watch detonator, which releases the pin from a grenade that's hidden in a pile of barrels. The resulting explosion rips a tiny hole in the ship's side, sufficient to put the stealth ship on the radar screen.

  The watch has a special detachable piece that Bond activates remotely using radio signals. First, he puts the remote unit with explosives in a jar. Then he pulls the pin from a grenade while making sure the spring-loaded spoon remains attached to the grenade. In this way, he can determine when the grenade will explode rather than having the grenade explode in his face. He slips the grenade into the jar. Later, Bond sends a signal from his wristwatch to the receiver inside the jar. The spring-loaded spoon is released and the grenade explodes.

  Also of interest is Wai Lin's wrist piton, not quite a watch but close enough. After all, Q could easily put a piton in any of Bond's wristwatches. In Carver's Hamburg headquarters, Wai Lin fires her wrist piton into the wall and then escapes by walking down the outside wall.
- *The World Is Not Enough*: Bond's Omega Seamaster wristwatch

has lasers and a tiny grappling hook. The blue night-light on the watch, common in today's world, enables Bond to see in the dark. And when chasing Renard down the Kazakhstan bunker, Bond aims the watch's grappling hook toward the ceiling and then leaps into action.

Today's real-life watches are far more sophisticated than the watches in Bond's universe. (That statement must be qualified slightly, as even today's watches don't include bullet-deflecting electromagnets and circular saws.) One example is the Casio Global Positioning System (GPS) wristwatch, which includes navigation features such as the latitude and longitude of the wearer's current position, a calendar, and an almanac. It uses satellite signals to display GPS locations.

The GlucoWatch Biographer wristwatch monitors the wearer's glucose without using needles. Instead, the watch uses tiny discs to extract glucose molecules from the extracellular fluid in the outer layer of the skin, so the discs don't need to break the skin. Small amounts of electricity collect the glucose, which reacts with glucose oxidase inside the disc. The result is hydrogen peroxide, which generates an electric signal that the watch converts into a glucose measurement.

As for the detonating, transmitting, and other activities performed by Bond's watches, these are standard methods found in Bond's pens, grenades, walking sticks, glasses, cameras, briefcases, and so forth. A detonator is a detonator, no matter where it is placed. A bomb is a bomb no matter where it is located. We'll discuss this subject in greater detail a little further on.

For now, we'll note briefly that today's watches can do a lot more than transmit simple detonate signals. A new wristwatch by IBM, for example, runs the Linux operating system. This means that Bond's watch could communicate wirelessly with computers anywhere, retrieving and sending data, photos, images, and statistics; processing information of any kind; and infiltrating and shutting down evil maniac computer systems all over the world. In addition, this means that Bond's watch could communicate with other wireless devices such as cell phones; send, retrieve, and display e-mail; and print out paper messages on printers. It could also be Bond's calendar, address book, and to-do list (1. Infiltrate Dr. No's hideout; 2. Set off bomb in Siberia; 3. Hijack a rocket into outer space; 4. Have martinis with Jill).

Other Devices That Go "Boom"

Besides wristwatches, a lot of other personal items used by Bond operate as bombs and explosive devices. For example, Bond often travels with exploding alarm clocks, as he does in *A Licence to Kill*. In that same movie, Bond's British passport doubles as a minibomb, which explodes when someone opens it. Also, Bond's toothpaste doubles as an explosive.

In *The World Is Not Enough*, Bond uses bagpipes as rifles and flamethrowers. In *Tomorrow Never Dies*, his cigarette lighter is a grenade. In *GoldenEye*, his pen is a grenade. And—dare we mention it—also in *GoldenEye*, his leg cast is a missile launcher.

Moving to *The Living Daylights*, we find Bond using a key ring as a bomb that he activates by blowing a whistle. In *The Man with the Golden Gun*, Bond's camera fires rockets. As a final example, *Thunderball* features air missiles with explosive heads.

An explosion is a violent release of energy. A loud noise, or at least a big pop, usually accompanies it. Along with the noise comes a pressure wave of gas, which decreases in pressure as it moves from the point of the explosion.

If something flammable explodes, then fireballs may also be released from the explosion. The fireballs ignite anything flammable encountered as they shoot outward.

As you might have guessed, more than one type of explosion exists. The first is known as a deflagrating explosion. In this case, material burns slowly; ceilings and walls may collapse, and sofas may sizzle and eventually be destroyed. Normal heat transfer causes the slow burn of a deflagrating explosion; thus the surrounding temperature and air pressure contribute significantly to the rate of explosion.

The second type of explosion, most often the type triggered by and encountered by Bond, is a detonating explosion. In this case, the burn rate is rapid, the energy release is high, and the pressure at the center of the explosion is also extremely high. Shock waves and associated pressure disperse the energy quickly, causing the explosion to spread quickly. The surrounding temperature and air pressure are not particularly significant to the rate of explosion.

Detonating explosives include dynamite, trinitrotoluene (TNT), nitroglycerine, mercury fulminate, and ammonium nitrate fuel oil. Once ignited, explosives burn quickly and produce huge amounts of hot gas. The gas expands and creates explosive pressure.

A logical way to deal with Bond's most peculiar exploding items is to lump them into categories:

- Minibombs: Exploding alarm clocks, passports, toothpaste, and key rings. These are probably detonating explosives.
- Strange missiles: Bagpipe rifles and flamethrowers, and leg-cast missile launchers.
- Grenades: Cigarette lighters and pens.

First we'll look at the minibombs. Most often, they are triggered by lifting a latch, pressing a key, opening a lid, or some other mechanism. Or they can be triggered by remote devices that transmit signals to the explosives.

Possibly, Bond's explosive material is nitroglycerin, or $C_3H_5N_3O_9$. When the carbon and hydrogen mix with oxygen, nitrogen is freed, causing the explosion. Nitroglycerin is a colorless liquid that is soluble in alcohol and insoluble in water.

In 1846, Ascanio Sobrero combined glycerol with nitric and sulfuric acids, creating nitroglycerin for the first time under laboratory conditions. Still, it was Alfred Nobel who first produced nitroglycerin for widespread use. He created a blasting cap, or igniter, which was essentially a wooden plug filled with gunpowder. By lighting a fuse, Nobel detonated the gunpowder, which then detonated the nitroglycerin.

Nobel created dynamite by mixing nitroglycerin with silica, which he then shaped into rods. He used the term *dynamite* for his new explosive paste, and he used his blasting caps to ignite it.[1]

Reviewing our list of Bond's minibombs—exploding alarm

clocks, passports, toothpaste, and key rings—it seems very likely that a form of dynamite or nitroglycerin would do the trick. We can certainly postulate that the exploding toothpaste was Nobel's dynamite paste. It's not much of a stretch to think of nitroglycerin at work in the alarm clocks, passports, and key rings, either.

As for the strange missiles—bagpipe rifles and flamethrowers, and leg-cast missile launchers—these are probably nothing more than (1) standard rifles and flamethrowers secured into bagpipes, and (2) standard launchers stuck inside leg casts. Nothing special is required to create these devices; a little ingenuity goes a long way.

Flamethrowers shoot burning fuel and hence spread fire. They've been around since the fifth century B.C., when they were big tubes filled with burning coal and sulfur. Men blew into one end of the tube to launch the flaming fuel. In World War I, the Germans used flamethrowers against the French and the British. The Germans used a small model, created by Richard Fiedler, which could be carried and used by one soldier. This Kleinflammenwerfer, as the Germans called it, used pressurized air with carbon dioxide or nitrogen, and it shot burning oil approximately 20 yards. If flamethrowers are big tubes that spew burning oil, why not disguise them as bagpipes?

And what about the grenades, or Bond's exploding cigarette lighters and pens?

Grenades, which are handheld and small, are detonated either on impact or by a timed fuse. While soldiers may prefer timed fuses, so they can race to safety before the grenades explode, Bond's cigarette-lighter grenades are probably the type that detonate on impact. We don't see Bond setting timers, nor do we see him pulling pins from his grenades before launching them. Rather, we see Bond debonairly lighting a cigarette, then throwing the lighter in the direction of an evildoer.

As for pens and pencils, even back in World War II, pencil fuses were used to detonate explosives. A spy pulled the safety pin from a timed pencil, then squeezed the pencil in a particular place, breaking an ampoule of acid inside the pencil. The acid then corroded a wire, which released a prong that banged against the detonator cap. These were, as you may have noted, timed grenades.

Whatever Bond is using to cause explosions—pens, pencils, leg casts, or bagpipes—there's one method he can always count on to get the job done: pulling the trigger of his Walther PPK.

5

Stopping Nuclear War

I never joke about my work, 007.
—Q TO JAMES BOND, *GOLDFINGER*

The earliest James Bond films appeared during the height of the cold war, so it's not surprising that nuclear weapons played a significant role in many plots. In *Dr. No*, the title villain worked from his own atomic power plant located on an isolated island in the Bahamas. While it served as an interesting location for the final confrontation between Bond and No, the power plant was fairly unimportant to the plot of the movie. There was no atomic angle at all in *From Russia with Love*. That was not the case with *Goldfinger*, the third Bond movie and the first to use a nuclear device as a weapon, and in a quite unconventional manner.

The Original Dirty Bomb

Goldfinger succeeded as an epic spy adventure because it not only featured a memorable hero in James Bond but also presented a thoroughly entertaining villain in Auric Goldfinger. The German actor Gert Fröbe transformed Goldfinger from an ordinary criminal mastermind into a diabolical master of menace, a genius obsessed with gold and willing to do anything to obtain more of it. The director, Guy Hamilton, knew that filmgoers loved the villain as much as, if not more than, the hero in an action movie. Goldfinger, who remained totally immoral and utterly remorseless throughout his encounter with Bond, was a villain so despicable that his death, even though it occurred after the climax of the movie, ended the film on a high note.

In the story, Bond is taken prisoner by the evil Goldfinger. The criminal genius is incredibly rich because of his vast holdings in gold bullion, but he dreams of being even richer. At a meeting with a group of U.S. crime lords at his ranch in Kentucky, Goldfinger explains his plan. He intends on robbing Fort Knox using various apparatuses the criminals have smuggled into the country for him. After revealing his plan to the crooks in a giant underground library, as well as to Bond, who is hiding in a ventilation shaft, Goldfinger, for some inexplicable reason, seals the criminals in his library and proceeds to kill them with poison gas. Goldfinger's confession leads us to wonder why he bothered wasting his time and energy to tell them the plan in the first place. We're forced to conclude from this bizarre behavior that Goldfinger is a megalomaniac with a compulsion to brag about his plans, even to people he plans to kill.

Our suspicions are soon confirmed when Bond confronts Goldfinger about the planned robbery of Fort Knox. The secret agent hero tells Goldfinger that such a crime is impossible. In 1964, $15 billion in gold was kept in Fort Knox, and the gold weighed more than 10,000 tons. Loading that much gold into trucks would take sixty men twelve days to pack the bars into two hundred trucks. At best, Goldfinger would only have a few hours after using poison gas to kill all the soldiers at the fort before new troops arrived.

A less arrogant criminal mastermind would tell Bond to mind his own business, but Goldfinger lets Bond ramble on and on. It doesn't take Bond long to realize that Goldfinger doesn't plan to actually *steal* the gold from Fort Knox. Instead, he plans to make it radioactive.

Goldfinger has obtained a nuclear device from the Red Chinese, the cold war nickname used to identify the mainland Chinese government. It's a "very dirty" bomb, Goldfinger remarks, made from cobalt and iodine. Bond remarks that, if exploded in the center of Fort Knox, the bomb will turn the entire gold supply of the United States radioactive for fifty-seven years. "Fifty-eight years," Goldfinger says, correcting Bond. The U.S. gold supply would thus become worthless. The free world would plunge into economic chaos, and the value of Goldfinger's gold hoard would multiply by a factor of ten.

Needless to say, after a bunch of heroics, a memorable battle with the villain Oddjob, and a countdown a bit too close for comfort, Bond succeeds in stopping the nuclear bomb from exploding. The gold in Fort Knox doesn't become radioactive, and the economy of the free world remains safe. But the plan still sounds good, and watching the movie you can't help wondering why no other megalomaniac villain tries it some other time—maybe they realize that the plan sounds better than it really is.

By definition, all nuclear explosives are dirty bombs. The term applies to any atomic bomb that generates a large amount of radioactive waste in the form of nuclear fallout. Fallout is the radioactive dust created when an atomic bomb explodes. In the fireball that forms within seconds after detonation, every type of material, including the ground and anything on it, vaporizes. When this material condenses in the mushroom cloud following the explosion, it turns into a light, sandlike material. This extremely radioactive dust falls to the earth, emitting deadly gamma rays in all directions. Due to the inefficiency of early nuclear weapons, they tended to disperse large amounts of unused fissionable material. By the time a significant percentage of the atoms have fissioned, their thermal kinetic energy is so high that the pit will expand enough to shut down the reaction in only microseconds. The practical efficiency limit of a typical pure fission bomb is about 25 percent, and it could be much less. Little Boy, the nuclear bomb dropped on Hiroshima, had an efficiency limit of only 1.4 percent.

In recent times, the term *dirty bomb* has been used to describe a bomb that would disperse radioactive fallout over a fairly wide area using conventional material. Some nuclear weapon designs feature the inclusion of a salting metal (most commonly cobalt) that will create large amounts of long-lasting fallout radiation when contaminated by the weapon core.

Despite a great deal of publicity and news devoted to dirty bombs, due no doubt to the possibility of terrorists using them, no dirty bomb has ever been detonated. Most weapons experts estimate that the damage done by such a bomb exploded on ground level would actually be minimal, as the radioactive particles would not scatter over a great distance. The main fear of a modern dirty bomb is the panic that would ensue from its use.

In *Goldfinger*, the dirty bomb that the villain planned to explode in the heart of Fort Knox was a fairly small nuclear device that included as a salting metal a mixture of cobalt and iodine. The fallout created from the cobalt-iodine mix was what Goldfinger counted on to make the gold in Fort Knox radioactive.

Neither Goldfinger nor James Bond seems to know much about atomic physics. Though a good amount of the gold in the vault would most likely be turned into gold leaf covering the walls, some of the gold would gain an extra neutron from the streams of subatomic particles let out by the blast and turn radioactive. The radioactive form of gold is extremely unstable, however, and it would turn to liquid mercury within a few days. So much for the gold becoming radioactive for fifty-eight years, a number no scientist has ever been able to explain. Regardless, Goldfinger's very dirty bomb would still wreck the economy of the free world and definitely make his gold a lot more valuable!

Nuking Miami Beach

The next Bond film, *Thunderball*, was a major box office hit and one of the most successful action films ever produced. As in *Goldfinger*, at its heart was a nuclear explosion. But this blast had nothing to do with the Chinese or turning gold radioactive. The bomb was British made and the menace was very personal. Instead of nuking the gold in Fort Knox, the villain of *Thunderball* intended to destroy the city of Miami Beach, Florida, with a nuclear weapon thirty times more powerful than the bomb dropped on Hiroshima.

In *Thunderball*, SPECTRE has a new mission. The criminal organization plans to steal two nuclear weapons from England and use them to blackmail the United States and Great Britain. Heading the operation is the number-two man at SPECTRE, Emilio Largo. Why two nuclear bombs? Because if the United States and Britain hesitate to pay the $100 million extortion fee, SPECTRE will detonate one of the missiles while keeping the other in reserve. Nothing makes reluctant politicians take notice of a threat like blowing up a major population center.

Largo steals the bombs by having one of his men, disguised by plastic surgery, hijack a British Vulcan delta-winged nuclear bomber carrying two Avro Blue Steel nuclear missiles. The faux pilot lands the plane off the coast of Nassau in the Bahamas. The plane and the two missiles are hidden beneath the water, kept there while Largo waits for payment. The target for the first bomb is Miami Beach.

The Blue Steel nuclear missile was developed in England in 1954 and was carried by British warplanes from 1961 to 1969. The missile was 35 feet long with a 13-foot wingspan. It weighed approximately 15,000 pounds and had a range of nearly 100 miles. It had a two-chamber rocket engine that used hydrogen peroxide and kerosene. The Blue Steel missile carried a 1-megaton Red Snow nuclear warhead. Normally, the rocket was fired from the Vulcan bomber while it was in the air. The missile flew at approximately 1,000 mph. As it drew close to its target, the rocket engine pushed its speed up to mach 3, almost 2,300 mph. When directly over the target, the missile's engine would cut off, and the missile would drop a few hundred feet before detonating in midair.[1]

We don't know exactly how Largo planned to detonate the Blue Steel nuclear missile in Miami Beach, since he was stopped by James Bond before he ever had a chance to try. Most likely, he planned to carry the bomb into Miami harbor, unload it onto a truck, park the truck in front of a hotel, and then detonate the Red Snow warhead when he was a safe distance away. Which raises a frightening question that is more applicable today than ever: how much damage could one nuclear bomb in the hands of criminals or terrorists do to a modern city?

In 1965, at the time of *Thunderball*, Miami Beach had a population of approximately 34,000. Miami, the city surrounding Miami

Beach, had a population of approximately 150,000. Miami-Dade County, in which Miami is located, had a population of approximately 1 million.

For comparison's sake, according to 2002 figures, Miami Beach covers 7 square miles and is one of the most densely populated cities in the United States. Miami is 34 square miles and is the eleventh largest populated area in the United States. Miami-Dade County has a population of over 2,253,000.[2]

A Blue Steel missile carried a 1-megaton Red Snow atomic warhead. A megaton is a unit of energy approximately equal to the energy released by the explosion of 1 million tons of dynamite. By comparison, the atomic bomb dropped on Hiroshima was 13 kilotons. A megaton atomic bomb is seventy-seven times more powerful than the bomb dropped on Hiroshima. A modern 1-megaton bomb weighs approximately 1,000 pounds, so it can easily be transported in an ordinary truck without problems.

Assuming Largo smuggled the 1-megaton Red Snow warhead into Miami Beach and then set it off, what would happen?

At detonation, the temperature at the bomb site would rise to 100 million degrees Celsius (nearly 200 million degrees Fahrenheit). The heat would create a fireball approximately 1 mile in diameter, which, within a few seconds, would drop to 11 million degrees. The flash created by the explosion and resulting fireball would light the sky for hundreds of miles. Fifty miles away, the blast would be brighter than the sun. People five miles away from the blast would suffer dangerous third-degree burns. At a distance of six miles, the heat would produce second-degree burns, and at seven miles, it would cause first-degree burns.

During the next ten minutes, a mushroom cloud would form over the blast site, rising more than 15 miles into the sky and billowing out to around 30 miles in diameter. At ground zero there would be a crater 200 feet deep and 1,000 feet in diameter. The rim of the crater would be 1,000 feet wide and would consist of highly radioactive dirt and debris from the explosion. For more than half a mile, nothing would be left standing. At 1.5 miles from ground zero, only the wreckage of buildings made with reinforced concrete would remain. Less than 2 percent of the population would remain alive.

The wind from the blast would blow outward from the center of the explosion at speeds over 180 mph. The air pressure created by the explosion would be so great that it would cause most buildings to implode and their ruins to be carried away by the tremendous winds. Nothing but foundations would be left. Eardrums would shatter from the pressure. Half the people in this area, 4 miles from ground zero, would be dead, with another 40 percent of the population injured.

At 5 miles away, most residences would be destroyed or badly damaged. The upper floors of office buildings and the people in them would be smashed to the street. Five percent of the population would be dead, with half of the rest injured. The city of Miami would lie in ruins, with more than half the population dead and nearly all the major buildings higher than a few stories destroyed.

Unfortunately, a nuclear blast also causes damage that can't be seen or measured. Immediately after a nuclear explosion, a great amount of dirt and debris is carried high into the air, forming the gigantic mushroom cloud. This material slowly drifts downwind and falls back to the earth, contaminating thousands of square miles beyond the immediate blast area.

Within seven days, areas within 30 miles of ground zero would be exposed to a lethal dose of fallout radiation. That would include everywhere from Fort Lauderdale in the north to Key Largo in the south. Death would occur within a few hours of exposure. It would be at least ten years before the radioactivity levels dropped low enough for this area to be considered safe.

Within 90 miles of ground zero, a circle that stretches past Palm Beach in the north, the entire Florida peninsula to the west and south of Miami, and the island of Cuba to the east, the fallout would still be dangerous enough to kill. Death would occur in a period from two days to two weeks. Within an area of 160 miles from ground zero, approximately north to Vero Beach, Florida, radioactivity wouldn't kill but would cause extensive nerve damage as well as damage to the digestive tract. Living in such an area would be like committing slow suicide. Largo's 1-megaton bomb would have turned southern Florida into a radioactive no-man's-land.

A nuclear explosion also fuels *Octopussy* (1983). The Soviet general Orlov wants the Americans out of Europe so the Soviet Union can invade Germany. So Orlov tricks Octopussy, the heroine of the movie and a circus owner, into smuggling a nuclear device into the Feldstadt American airbase in Germany. The general plans to set off the bomb, killing thousands in a nuclear "accident." Orlov figures such an action would make Americans so unpopular they'd close all their European bases, making it impossible for them to stop a Soviet invasion afterward. Of course, James Bond stops the bomb from exploding in what has to be the most convoluted and ridiculous plot in the entire series. Again, the movie's producers seem unaware of exactly how much destruction would have resulted from even this "small" nuclear device.

Though produced more than a decade apart, *Thunderball* and *Octopussy* both make it quite clear why letting terrorists get their hands on nuclear weapons would be a very bad idea, then as well as now.

James Bond Meets Dr. Strangelove

After saving Miami Beach from nuclear disaster, there seemed little left for James Bond to do other than save the entire world from the same disaster. In the fifth movie of the series, *You Only Live Twice*, Bond battled SPECTRE again, but on a much larger scale. The villain in this adventure was number one, Ernst Stavro Blofeld, the leader of SPECTRE, working for the mainland Chinese government. His mission was to bring about a nuclear war between the United States and the Soviet Union.

Though the film features a number of highly entertaining segments capped by an unmatched battle sequence between Japanese Secret Service Ninjas and SPECTRE agents inside the interior of an extinct volcano, the plot often borders on the ridiculous. Someone is stealing manned space capsules in space. A team of U.S. astronauts is taken while in orbit, and then a team of Soviet cosmonauts is captured. Each country blames the other and the threats of violence escalate. Only Great Britain seems to understand that a third power might be involved. Their tracking stations indicate that the mysterious kidnapping rocket appears to come from Japan (see chapter 8, especially the section titled "The *You Only Live Twice* Gobbling Spaceship"). Anxious to prevent a full-scale nuclear war from breaking out between the Soviets and the Americans, the British secret service sends James Bond to investigate. Seeking to keep Bond's mission a secret, M arranges for Bond to be "killed" and then brought back to life undercover.

Roald Dahl's script borrowed the title from Ian Fleming's last Bond novel (Fleming began *On Her Majesty's Secret Service* but died before completing it) but little else. Despite cold war tensions, the notion that the United States and the Soviet Union could be duped into a nuclear war was about as believable as Blofeld being able to construct a massive spaceship based inside an extinct volcano in Japan with no one noticing.

Still, *You Only Live Twice* dealt with a subject that was extremely important to most Americans. In the late 1960s, the United States and the Soviet Union appeared set on a collision course that could only end in nuclear war. The one factor keeping either nation from acting was a nightmarish battle scenario known as mutual assured destruction (MAD). However, as pointed out only a few years before in the black comedy *Dr. Strangelove, or How I Learned to Stop Worrying and Love the Bomb*, as well as in *You Only Live Twice*, MAD worked only if the leaders of both nations were interested in maintaining peace and not waging war.

MAD states that a full-scale nuclear attack by one of the two major atomic powers (the United States and the Soviet Union) would result in the total annihilation of both the attacker and the defender. The theory assumes that each country has enough atomic weapons to destroy the other country, and that if either side was attacked for any reason by the other, it would retaliate with the same or greater force. In simple terms, MAD states that a nuclear attack would result in global disaster for both participants.

As we mentioned earlier in this book, MAD was the basis of

American (and later, Soviet) nuclear policy from the late 1950s through the late 1990s. Though the Bush administration calls MAD outdated, the U.S. government still tacitly accepts the basic concept of the theory.

Since neither the United States nor the Soviet Union was viewed as insane enough to risk its own destruction, it was assumed that neither country would launch a first-strike missile attack on the other. Belief in MAD kept the peace by threatening the total devastation of war.

The major problem with MAD was that enough weapons had to survive in the attacked country for it to respond with equal force. To ensure that this scenario held true, thousands of nuclear missiles were built, nuclear bomber forces were kept on high alert, and huge concrete underground missile silos were constructed in hopes of withstanding an atomic attack. Still, it wasn't until the development of submarines carrying ballistic missiles that the two major powers felt that MAD conditions were in place. Neither country was sure that after it launched a first-strike attack it could take out all the nuclear submarines hiding in the ocean before the other country retaliated with atomic weapons.

Still, even James Bond had his doubts about the effectiveness of MAD. Desperate men often commit desperate crimes. In *Octopussy*, mentioned in the previous section, General Orlov felt that the Soviet Union was losing the cold war to the United States. Rather than surrender his power, the general planned to detonate an atomic bomb in West Germany and blame the explosion on the U.S. Air Force. He was certain that such an event would drive the United States from Europe. Afterward, he planned an invasion, certain that the United States would not risk nuclear destruction to save Europe from communism.

Critics of MAD worried about scenarios like that shown in *You Only Live Twice*. MAD depended on the notion that no rogue state possessed nuclear weapons, or if it did, that it would never use them, because of the immediate global devastation. MAD worked only if the leaders of all nations were rational. Saddam Hussein and Kim Jong-il serve as strong arguments against trusting MAD too much.

Still, the Chinese government hired Blofeld to destroy the stability of MAD in *You Only Live Twice*. It was SPECTRE's goal to bring about war between the two nuclear powers, supposedly making it possible for the Chinese to gain control of the postapocalyptic world. The plot of *The Spy Who Loved Me* is based on the same scenario, except the madman villain is stealing nuclear submarines and hopes to start an underwater utopia. There was one interesting question never really covered by MAD: If the two superpowers destroyed each other in a nuclear war, would the rest of the world survive?

At the time, that was a question to which no one knew the answer. Many scientists warned that radioactive fallout from a nuclear war would kill everyone on Earth (as shown in Nevil Shute's 1957 novel *On the Beach*), while military leaders preached that fallout shelters would keep civilization going. Some people actually did believe that China, with its huge population and gigantic size, could survive a nuclear war—at least they did until the theory of nuclear winter was first advanced.

A famous effort to predict whether the effects of a large-scale nuclear war would bring about a nuclear winter was the 1983 TTAPS study (from the initials of the last names of its authors Richard P. Turco, Owen B. Toon, Thomas P. Ackerman, James B. Pollack, and Carl Sagan). The authors were inspired to write the paper by pondering the cooling effects of dust storms on Mars. To carry out a calculation of the effect, they used a simplified two-dimensional model of Earth's atmosphere that assumed conditions at a given latitude were constant. Currently, the consensus with more sophisticated calculations is that the model used in the TTAPS study probably overestimates the degree of cooling in its conclusions, though the amount of this overestimation remains unclear. Although such nuclear war would undoubtedly be devastating, the degree of damage to life on Earth as a whole still remains controversial.

As with many scientific theories, nuclear winter has become a debating point between scientists and politicians. Right-wing nationalists in the United States argue that nuclear winter is an unproven, alarmist theory aimed at leaving the United States help-less against its brutal enemies. Left-leaning liberals and many scientists counter that nuclear winter is a well-constructed theory that needs to be understood now, before any of the problems leading to it take place. They believe that once the damage of nuclear winter is done, there's no way to reverse the trend and make Earth healthy again.

The basic concept of nuclear winter is that a large-scale nuclear war involving the detonation of hundreds (if not thousands) of atomic bombs would change Earth's climate. It's thought that severely cold weather would be caused by the destruction of flammable targets located in major cities. These targets would be factories and industrial parks, whose destruction would send large amounts of smoke and soot into the stratosphere.

This new layer of particles would reduce significantly the amount of sunlight that would reach Earth's surface. These particles could remain in the stratosphere for weeks, months, or even years. Smoke from burning petroleum fuels and plastics absorbs sunlight much more than smoke from burning wood. The smoke and soot would be pushed by strong west-to-east winds, forming a uniform belt of particles encircling the Northern Hemisphere from 30° to 60° latitude. These thick black clouds could block much of the sun's light for periods as long as several weeks at a time, causing surface temperatures to drop by as much as 20°C (68°F). The darkness, killing frosts, and high doses of radiation from nuclear fallout would severely damage plant life, making agriculture near impossible. Add the effects of nuclear winter to the massive destruction of most civilization by atomic weapons, and the few survivors of the bombs would most likely die from starvation, exposure, and disease.

The Federal Emergency Management Agency (FEMA) has studied several nuclear winter scenarios. A typical nuclear attack would kill 86 million people and seriously injure 34 million. With only 2 million hospital beds available in the United States and Canada (and many of them in high-risk areas), the chance of those 34 million surviving is slim.[3]

The darkness caused by nuclear winter would hamper all activity for those who did survive. Cold temperatures would cause exposure, hypothermia, and frostbite. In severely cold conditions, wounds wouldn't heal well unless closed properly. Medical services would be almost nonexistent.[4]

Radioactivity in the air would be dangerous, as would be the effects of toxins in the air. Both would make normal healing almost impossible. Many common diseases controlled by antibiotics, including measles, whooping cough, meningitis, and scarlet fever, would escalate. Even bubonic plague might reappear.

Still worse, after the dust and soot in the atmosphere finally settle, the sun's rays would once again reach the Earth's surface. Most likely, the many atomic explosions would have destroyed much of the ozone layer in the stratosphere. Large amounts of ultraviolet light would reach the surface, eventually causing blindness in most humans and animals.[5]

Nuclear winter would affect the entire planet. If the predictions are true, China's leaders would be insane to hire Blofeld to provoke a nuclear war between the Soviet Union and the United States. Equally, Blofeld would be insane, no matter what he was promised for bringing about the apocalypse, to accept such a job. And Blofeld definitely is *not* insane.

Nuclear winter, like global warming, remains a scientific hot topic, divided as much by politics as by science. We can only hope that until a consensus is reached on both of these possible worldwide disasters, James Bond will be on call to save Earth whenever necessary.

6

Using Your Senses
Assorted Body Equipment

No well-dressed man should be without one.
—JAMES BOND, COMMENTING ON THE JETPACK, *THUNDERBALL*

Any modern secret agent needs good equipment. James Bond has more than his fair share of items he can call on in an emergency. Many of these gadgets are seen only once or twice in the series, but they all serve one important function: they help keep Bond alive when otherwise he'd be dead. The following are just a few of the important devices Q Division has come up with over the years that have some basis in real science.

The Jetpack

According to a very unscientific poll conducted in 2003 by Carling Extra Cold Beer, 3,000 film fans named the jetpack in *Thunderball* (1965) as the best gadget ever to appear in a James Bond movie. It's a fact that probably would have given Wendell F. Moore more than a measure of satisfaction if he hadn't died in 1969. Unlike most of the unusual gadgets and body gear used by James Bond over the years, the personal jetpack really existed. Nor was it created merely as a novelty device for a Bond film. Instead, it was designed for use by the U.S. Army in wartime. However, it was James Bond who made it famous.

The jetpack appears in the opening sequence of *Thunderball*. After a lethal exchange with an old enemy, Bond escapes the villain's associates by donning a jetpack and helmet and shooting high into the air. It is one of the most memorable scenes ever to open a movie and by far the most entertaining.

In the late 1950s, the U.S. Army wanted a practical machine for improving troop mobility. In 1960, the Army Transportation Command awarded Bell Aerosystems a $150,000 contract to develop a Small Rocket Lifting Device (SRLD). At the time, Moore was a scientist who worked for Bell Aerosystems in Buffalo, New York. He had previously been a member of the design team that built the most famous series of Bell experimental rocket air-planes, including the Bell X-1, which broke the sound barrier in 1947.

Moore dealt with small rockets fueled by hydrogen peroxide. According to a number of interviews he gave over the years, Moore came up with the idea of a man flying by the use of small rockets on his back one evening while doodling. Funded by the army grant, Moore turned his doodle into an actual rocket belt. The invention was little more than a high-powered rocket strapped to a man's back. The jetpack used pressure from liquid nitrogen to force hydrogen peroxide into a catalyst chamber, where it reacted with silver screens coated with samarium nitrate. The mix created a jet of hot, high-pressured steam that provided the thrust that lifted the user into the air.

On April 20, 1961, Harold Graham, an associate of Moore's at Bell Aerosystems, flew 112 feet outdoors using the rocket belt. The entire flight lasted 13 seconds. Unfortunately, the Bell jetpack was highly impractical. One wrong move and the pilot was badly burned by the steam. Equally dangerous, the flier had to use his own legs as landing gear. In addition, the jetpack made an incredibly loud noise when in operation.

Despite all of its flaws, the Bell jetpack fascinated the public. The device was demonstrated numerous times around the world and on television and at air shows. It was exhibited at the 1964 New York World's Fair and appeared in the half-time celebration of the first Super Bowl. The SRLD's pilot for all these appearances was a neighbor of Moore's, William P. Suitor. When the opportunity arose to use the SRLD in *Thunderball*, the film's producers jumped at the chance.

The army never used the Bell jetpack for the simple reason that it could carry enough fuel for only a 20-second flight. When Moore died in 1969, the jetpack was retired. It returned for a flight at the opening of the 1984 Olympics in Los Angeles.

The concept of a personal flying device refused to die. An August 2000 news release described the Solo Trek XFV, made by Millennium Jet, Inc., as a vertical one-man jet that could fly up to 80 mph and for 150 miles before refueling. Unfortunately, the Solo Trek never lived up to its potential or its costs.

In December 2003, Trek Aerospace, the company that designed the Solo Trek XFV, offered a working prototype of the flying machine for auction on eBay. The minimum bid was $1 million, and that didn't include any rights to actually fly the machine!

According to Michael Moshier, the chief executive of Trek Aerospace, the small company spent seven years and around $4 million (including funds from a military grant) to design the personal flying machine. Unfortunately, the company ran out of cash needed for further research, which was why they put the one-of-a-kind prototype for sale on eBay in hopes of raising more money. It never sold.

The flying machine stands about 7 feet high and weighs slightly more than 300 pounds. It has two overhead engines latched above the tripod frame that holds the pilot. The Solo Trek can carry a total weight of 240 pounds. The pilot steps into the middle of the frame, straps it to his body in a standing position, and then uses two joysticks to control the device.

The Solo Trek takes off and lands vertically. It flies at speeds up to 69 mph, and it has a range of 100 miles. The Solo Trek flew for the first time in December 2001.

Moshier says the winner of the eBay auction will have to sign an agreement not to operate the flying machine. His company fears a lawsuit if someone is injured. Put a few guns on it, maybe a missile or two, and who knows: it might just pop up in James Bond number 21!

Voice Box

Diamonds Are Forever (1971) was the seventh James Bond film and the sixth and final time that Sean Connery played the superspy for Eon Productions. Connery did return as Bond one last time, in *Never Say Never Again* (1983), a remake of *Thunderball* that was, in a word, unfortunate. Connery had left the Bond series after *You Only Live Twice*. After the tepid reception to George Lazenby in *On Her Majesty's Secret Service*, however, Connery was lured back to the series by the producer Cubby Broccoli with a substantial raise, making him the highest-paid actor in the world at the time. The movie was a hit, making just under $44 million in the United States and $116 million in worldwide revenue, nearly double what the Lazenby Bond film earned but still down sharply from what *You Only Live Twice* made four years earlier.[1]

The film's failure to attract the size of audience that had flocked to see the earlier Connery Bond movies was due in no small measure to an erratic and sometimes idiotic plot that borrowed its name and basic theme, diamond smuggling, from Ian Fleming's novel, but little else. The story begins, after a brief prologue in which Bond tracks down and seemingly kills Ernst Stavro Blofeld, with the spy infiltrating a diamond-smuggling ring. The story staggers from Amsterdam to Las Vegas, with the apparent villain of the piece being the reclusive billionaire Willard Whyte. Much in the vein of Howard Hughes, Whyte lives on top of a Las Vegas skyscraper, never seen by the public. He commands his underlings by intercom or telephone. Only late in the film does Bond finally uncover the truth. Whyte's a prisoner of Blofeld (who has several doubles working for him), who is using a voice synthesis machine to match the billionaire's voice. The movie closes with an impossible space laser that uses diamonds to focus energy beams, and with Bond killing Blofeld yet again.

In 1971, the voice synthesizer was an amazing device that Blofeld merely held to his larynx to change his voice into that of Whyte's. Bond used a similar device later in the film, Q having developed one as a Christmas present for his children. The handheld device was offered without any scientific explanation or double-talk. A good idea, since at the time such devices were impossible.

More than thirty years after *Diamonds Are Forever* was released, the technology shown in the film is just now starting to appear. And, needless to say, it needs a computer to make it work.

Speech recognition software was originally developed for computers primarily to perform simple commands as seen on the *Star Trek* TV show and science-fiction movies. Later, programs were written to handle basic word processing. The original versions of these programs were notoriously unreliable. Nearly forty years of updates, however, have improved the accuracy of the software. Most speech recognition programs are quite reliable, though they still need to be taught to recognize their user's voice and accent.[2]

Going in the other direction, having computers convert text into speech has proven much more difficult. The operation is known as text-to-speech (TTS) synthesis. A major breakthrough in TTS occurred in 1976 when Ray Kurzweil introduced his Reading Machine, which could scan printed copy and speak it. Following Kurzweil's program came Voice XML from Bell Labs, based on software developed for customer and call center use. Motorola developed VoxML in 1998, also for interactive telephone use, emphasizing voice recognition and response instead of keypad sounds. Until recently, digital speech technology was extremely expensive, but with improvements in digital technology, prices have dropped so much that speech synthesis and recognition software are available to anyone who owns a computer and a microphone.[3]

In the past, synthesized voices sounded mechanical and unnatural. Today, speech synthesis sounds more normal due to the wide choice of voice types, cadence, and accents. Numerous companies offer programs where you speak into the PC's microphone and the new voice is heard on your computer speakers or on the telephone at the other end of the line. Unlike traditional voice changers that are hardware-based products, these programs are innovative software products running on your PC.

Typical of these software programs is Virtual Personality, described as "a professional telephone-Internet voice changer utility." You speak into the PC microphone, and your natural voice is digitally sampled by the multimedia components already installed in your PC. It is then modified by a sophisticated algorithm inside Virtual Personality and applied to the PC's speakers' output for real-time use. The software can change your voice to mimic any from a cast of twenty-four different personalities, representing voice sources of boys, girls, men, or women. You can also adjust voice pitch and timbre levels to make different sex, age, and voice characteristics.[4]

In other words, while *Diamonds Are Forever* was one of the least interesting Bond films, its prediction of a voice-changing machine was right on the mark—just thirty years too soon!

Fake Fingerprints

In *Diamonds Are Forever*, James Bond is sent by M to break up a diamond-smuggling operation headquartered in Amsterdam. Bond meets the beautiful Tiffany Case in the course of his investigations and tells her that he is Peter Franks, a notorious thief. Unbeknownst to James, Tiffany obtains a set of his fingerprints and compares them to the fingerprints on file of the real Franks. The prints are a perfect match. Knowing that fingerprints are the most reliable identification method available in the early 1970s, Tiffany feels confident that Bond and Franks are the same man. Only later, during a phone conversation with Q, do we discover that Bond is wearing thin latex membranes on his fingers that duplicate the other man's prints. "An obvious little notion," Q calls them. Obvious or impossible?

Fingerprints are one of the cornerstones of modern forensic science and have been used to verify a person's identity for nearly 100 years. They are routinely used by police to link a suspect to a crime. Even when the police don't have a suspect, fingerprints are important to almost every criminal investigation. Prints can provide clues about the criminal's size, gender, and sometimes occupation. Small prints tend to be made by small people, and prints on a wall usually indicate the suspect's height. Prints can prove or disprove the alibi of a suspect. Even the absence of prints may be a key factor. Suicide scenes, obviously, are not usually wiped clean of prints.

Fingerprints come into being several months before a baby is born, when ridges form on the skin of the baby's fingers and thumbs. These ridges arrange themselves in more or less regular patterns. For purposes of classification, experts divide these ridge patterns into three basic classes: arches, loops, and whorls. Each class can be further divided into numerous subcategories.[5]

The process of analyzing fingerprints is known in the United States as dactylography. Strangely enough, although fingerprints are most often associated with the FBI and U.S. police departments, they were first used in India. In 1858, a British magistrate named William Herschel, who worked in a village in Bengal, India, realized that every fingerprint was unique and unchanging. In India, there was a tradition, especially popular among Bengalis, of using a fingerprint as a signature. It was Herschel, however, who first used the uniqueness of such prints to resolve a contract dispute.

Herschel was the first person to take the fingerprints of all fingers of a criminal suspect, but it was Juan Vucetich, an Argentinian police official, who standardized the practice approximately thirty years later. A few years after that, Francis Galton, a cousin of Charles Darwin and a genius in many fields, used mathematics to prove that fingerprints were unique, giving them scientific credibility. Galton identified common patterns in fingerprints and helped devise a classification system for them. He also estimated that the probability of two people having the same fingerprint was 1 in 64 billion. Recent mathematical studies suggest even higher odds.

In 1891, Edward Henry, a Bengali police inspector who later became the assistant commissioner of the London metropolitan police, expanded Herschel's fingerprinting techniques, and using Galton's theories, fully developed a system of fingerprint classification. In 1900, Scotland Yard abandoned the French fingerprint system it was using and began using the Galton-Henry system. This system became the FBI–National Crime Information Center system. The first criminal trial to result in a conviction based on fingerprint evidence took place in 1901.

The Galton-Henry system stated that every fingerprint has certain ridge patterns. Along with these ridge patterns, there are minute variations and irregularities in the ridges, described as ridge characteristics. Examples of ridge characteristics are bifurcations, ridge dots, and crossings. Every print has unique pattern and ridge characteristics. No two prints are exactly the same—even the prints of identical twins differ. Prints remain the same throughout a person's life. Only by use of powerful acid or fire can fingerprints be altered.

Fingerprinting was used not only to identify criminals in most countries but also to register foreigners, Jews, and other ethnic groups. Opposition to ethnic fingerprinting was an important part of Mohandas Gandhi's peace movement in India, as well as the anti-Apartheid movement in South Africa, because of the strong association of fingerprints with criminals.[6]

When prints are found, an expert compares them with samples from a suspect. The criminologist first compares ridge patterns and then looks for ridge characteristics. When these match, they are known as points of comparison. In the United States, prints must match by twelve points of comparison before an identification is considered positive. The FBI keeps millions of prints on file, and most police departments keep files of unidentified prints from unsolved crimes in the hope that someday a match will be found.

Fingerprints fall into two categories: visible and latent. Visible prints are ones that can be seen and are usually made by blood, dirt, or grease. Latent prints are usually invisible to the naked eye and must be developed before they can be seen and photographed. Latent prints are usually discovered through dusting. Most people's fingers have a layer of oil and sweat. When fingers touch a smooth surface, the friction created releases the oil from between the fingers' ridges. When powder is applied to the surface, it sticks to the oil and brings out the pattern. Dusting relies on a vast number of different powders to bring out fingerprints on surfaces. It's the one area of police work most people are familiar with from seeing it done on innumerable TV shows.

Dusting is the most frequently used method for developing latent fingerprints, but there are other methods of detecting prints as well. With iodine fuming, iodine crystals are put into a pipe. The fumes are then blown onto the surface where the prints may exist. Ninhydrin spray is another method particularly useful in locating old prints. Ninhydrin will reveal prints made over thirty years earlier. Silver nitrate is a third method of locating latent prints. It can be painted or sprayed onto a surface and left to dry. The chemical bonds with the salts in perspiration.[7]

In the 1970s, after being in use for nearly three-quarters of a century, fingerprinting was considered the most accurate method of verifying a person's identity. Thus, it wasn't surprising that Tiffany Case was certain that James Bond was actually Peter Franks. While the entire fingerprint episode was hardly noticed in reviews of the film, it actually depicted a major scientific breakthrough that wasn't made until thirty years later.

In those days, criminals wore gloves to avoid leaving fingerprints at the crime scene. And even that wasn't always foolproof, since plastic gloves retained fingerprints. Crooks had to be extremely careful, or had to know the engineering professor Tsutomu Matsumoto of the Graduate School of Environment and Information Sciences at the University of Yokohama, Japan.

Professor Matsumoto and his colleagues at the University of Yokohama set out to check the reliability of fingerprint scanners. Their first experiment used phony fingers made by pressing a finger into malleable plastic used by model makers, then filling the indentations of the mold with gelatin. These Jell-O–style fingers deceived fingerprint readers 80 percent of the time.[8]

In a second experiment, Dr. Matsumoto created fake fingerprints using prints left on a piece of glass. As a first step, the scientists made a hardened image of the print using a glue that stuck to the ridges left behind when a finger touches a hard surface. The enhanced print was then photographed by a digital camera and sharpened on a computer using Adobe Photoshop to emphasize the differences between the ridges and the gaps. This new print was transferred to a photosensitive page and then etched into copper, turning a two-dimensional picture into a three-dimensional mold. As in the first experiment, the fake fingers created from the mold fooled fingerprint readers 80 percent of the time.[9]

Dr. Matsumoto and his associates presented their work in January 2002 at the electronic imaging conference sponsored by the International Society for Optical Engineering.

Still, 80 percent isn't the best of odds when you're a spy. Four out of five won't do if being number five means being killed. For those who want even better results, we offer Starbug and Lisa.

Starbug and Lisa are two German hackers who say they've developed a technique to fool fingerprint scanners used to authenticate electronic purchasing systems. The system creates latex fingertip patches designed to be used while under observation. James Bond, anyone?

The two hackers displayed their method in August 2003 at the Chaos Computer Camp, an open-air event held in Berlin. The technique uses graphite powder and adhesive tape to lift fingerprints off surfaces. Next, a digital picture of the fingerprint is taken, enhanced by software, printed onto foil, and then transferred to a photosensitive printed circuit board. The board is exposed and etched to create the three-dimensional structure of the fingerprint. The final image is transferred to liquid latex. The thin latex piece is then attached to the user's fingertip.[10]

According to Starbug, the most difficult part of the process is lifting the dried latex material in a sheet thin enough to be relatively invisible to observers. Starbug says that many sophisticated fingerprint-scanning systems also measure pulse and perspiration on the fingertip. But latex is thin enough to allow such information to be detected through the material. In essence, the hackers' system is quite similar to the one developed by Dr. Matsumoto.

Starbug and Lisa say that they produced artificial fingerprints to point out how easy it is to fool biometric scanners. The United States is pushing for European nations to include fingerprints, iris scans, or face recognition data on all passports by the end of 2006. According to Starbug, his artificial fingerprints could be used at border crossings without any difficulty. Starbug and Lisa claim they've also developed computer techniques for defeating face recognition, iris scan, and voice-print biometrics.[11]

During the past several years, biometric scanning devices that connect directly to home computers and identify users' fingerprints have become commonplace. At least two dozen different companies offer computer plug-ins that digitally verify thumbprints in seconds. As of this writing, the Jewel Food Store chain in Chicago is actually offering customers the option of using a thumbprint scan instead of their credit cards to charge food.

It seems as if the real world is finally catching up with the world of James Bond.

7

Getting Away from It All
In the Air and on the Sea

*Little Nellie got a hot reception. Four big shots made
improper advances toward her, but she defended her honor
with great success.*
—007 IN A MESSAGE TO Q, *YOU ONLY LIVE TWICE*

In the 1930s and 1940s, secret agents were more secret than agents. The world was a big place, and spies didn't travel much. Mostly, they just spied, trying to learn important enemy secrets and then somehow relaying that information back to headquarters. That's not the case in the modern world of James Bond. Spies are always on the move, changing countries as quickly as some people change clothing.

James Bond has a well-earned reputation for traveling around the world during the course of his adventures. His ability to function intelligently under the most dire circumstances serves him well when he's riding a motorcycle through the streets of Hong Kong, flying a hang glider in the Bahamas, or bungee jumping down the side of a dam in the Soviet Union.

In this chapter, we examine some of Commander Bond's more interesting adventures in the air and on the sea.

Goldfinger out the Window

Ask baby boomers what happens if you shoot a handgun in an airplane flying at 35,000 feet and the bullet blows out a window. Most likely, they'll relate a gruesome story of passengers spinning around the cabin like crazed balloons until they're sucked through the hole like carpenter's putty being squeezed through a crack in the wall. That's because they belong to the "Goldfinger generation," the millions of people who saw *Goldfinger* in the movies or on television and because of it believe one of the most enduring urban legends of all time.

In the climactic scene of *Goldfinger*, James Bond and the evil Auric Goldfinger stand face to face for the last time, in the main cabin of a jet flying at 35,000 feet. Goldfinger has a gun and intends to kill Bond. Earlier in the film, Bond lectures one of the other bad guys about why firing a gun in the pressurized cabin of an airplane is a dangerous idea: a bullet might puncture the skin of the plane, making a hole leading from the highly pressurized air of the cabin to the low-pressure air outside. According to Bond, the two pressures would try to equalize, creating a whirlwind of high-pressure air inside the cabin as it whips through the hole to the lower pressure outside. Anything not buckled down inside the cabin, whatever its size, would be caught in this minitornado and sucked out the hole into the atmosphere.

Bond warns Goldfinger, but the villain doesn't care. The two men fight, and Goldfinger fires the gun. A stray bullet hits one of the cabin windows and shatters it. The pressurized air in the cabin rushes through the cabin just as predicted. Bond manages to grab a doorframe; Goldfinger isn't so lucky. He flies around inside the plane for an instant, like a dust mote caught in a sandstorm, then he's sucked against the tiny square window and pulled through it. His body melts like soft ice cream being slurped through a straw. It's a very unpleasant way to die. When Pussy Galore, the plane's pilot, finally manages to land the plane, there's no sign of the four-inch-by four-inch- by twenty-foot-long corpse of Goldfinger. But the memory of him being squeezed like toothpaste is one that remains with moviegoers forever.

This scene is the basis for the second scientific urban legend that began with *Goldfinger* (the first was covered in chapter 5). Again, the moviemakers aren't at fault, as Ian Fleming described the scene in great detail in his novel. Though, in the original story, Oddjob was killed by the decompression. As in the movie, the book sets up the scene earlier in the story with Bond explaining on a plane how a bullet hole would cause a disaster. Then at the end of the story, as any good author would, Fleming offers a demonstration of the event actually taking place.

How widespread is this urban myth? Evidence from newspapers around the world during the past several years shows it is still believed by people in all of the major industrialized countries. With the threat of airline hijackings once again in the news, several countries have proposed putting armed sky marshals on all flights. The thought of having someone with a gun in a pressurized cabin is enough to cause nightmares.

Testifying at a hearing about putting armed sky marshals on British flights, a British expert on airline safety declared that a bullet destroying an aircraft window in real life would have dire consequences for nearby passengers, as dramatized in *Goldfinger*. According to the expert, as air rushes out to equalize the pressure differential between inside and out, it can reach the speed of sound and pick up people and objects.[1]

In the letters page of the *Arizona Republic*, after someone suggested that if most passengers carried guns on planes, flights would be much safer, a reader sent the following comment: "Good grief! Hasn't she ever heard of explosive decompression? Anyone with a gram of gray matter knows that a pressurized aircraft cabin at 30,000 feet is the worst possible place to discharge a firearm."[2]

The dramatic phrase "explosive decompression" sent us scrambling for an expert on airplane cabins. An interview with Robert Horsky, a ground crew mechanic for United Airlines, provided us with the reassuring truth.

An airplane flies at approximately 30,000 feet. The air pressure at that altitude, 4.3 pounds per square inch (psi), is much lower than at sea level, 14.7 psi. The inside of the aircraft is kept at a constant 8.5 psi. The cabin is pressurized by a constant flow of compressed air from the jet engines, which flows through the air conditioning and pneumatic systems. Inside the aircraft are two "holes" used to regulate the cabin pressurization. They're called the outflow valves, one located in the front of the plane and the other at the rear. Their function is to modulate and maintain a desired cabin pressure that is kept constant when flying at certain altitudes. This operation is performed automatically.

There are other tiny holes where pressurized air leaks from a plane; hence, it's impossible to fully seal the plane. The cabin is constantly being pressurized and recharged by the bleed air from the engines.

The effect of a gun being fired in an airplane is entirely dependent on the size of the hole caused by the bullet. If the bullet passes straight through a window, creating a 1-inch hole, it wouldn't cause rapid (or explosive) decompression. Cabin air would whistle out the hole, but the outflow valves would automatically respond to the sudden loss of air by closing the valves a little to compensate for the air leak. If a window is somehow blown out of the cabin by a shooting, the decompression would be more serious, but the outflow valves would still be able to keep the cabin pressure fairly steady, giving the pilot enough time to bring the craft down to a safer altitude.

If the hole resulting from the firearm discharge is big, the cabin could depressurize rapidly. As with a balloon, the bigger the hole, the faster the air would leak out. A hole the size of a finger would not have a significant effect on a big commercial aircraft.[3]

Therefore, Horsky assured us, the *Goldfinger* scenario was totally fiction. As a final measure of reassurance, he related the dramatic story of a Boeing 737 flying to Hawaii, where the front roof section of the first-class cabin ripped off at 24,000 feet due to undetected cracks in the airframe. In an instant, the passengers found themselves flying in a plane with an 18-foot hole in the roof. A flight attendant on her feet was swept from the cabin due to the speed and descent of the jet. Fortunately, the cockpit was undamaged, and the airplane landed safely with no other casualties.[4]

Goldfinger propelled James Bond to international stardom. Its combination of fast action, beautiful women, evil villains, and super-science made the movie a monster hit and established the pattern for all Bond pictures to follow. It also stands unique as quite likely the only movie in history to establish two urban legends that have remained part of our culture for more than forty years. This just proves how popular and pervasive the James Bond phenomenon has become in our lives.

Little Nellie—Autogyro

The prologue to *You Only Live Twice* establishes the theme for the rest of the movie. A mysterious unmarked spaceship is kidnapping manned U.S. and Soviet spaceships in orbit, causing the two megapower nations to act like little children and start making threats about all-out nuclear war if their astronauts are not returned. In the meantime, in Hong Kong, the British agent James Bond is caught with his pants down in a compromising situation and is shot to death by vengeful mob figures. The death, of course, is merely a ruse to permit a now "dead" Bond to investigate the fate of the missing spaceships. He has to act carefully, because every secret agent knows you only live twice.

In the course of his investigations, Bond asks Tiger Tanaka, the head of the Japanese secret service, to contact M and arrange for Little Nellie to be delivered along with her father. Nellie arrives disassembled in four large suitcases, with her father, Ordinance Master Q, close behind. Little Nellie, when fully assembled, is a real aircraft, a Wallis WA-116 autogyro. The flying machine is equipped with two fixed forward-firing machine guns synchronized to 100 yards, using incendiary bullets and high explosives; two forward-firing rocket launchers on either side; a group of heat-seeking air-to-air missiles; two flamethrowers, range approximately 80 yards; two smoke ejectors; some aerial mines; and a camera and radio built into the pilot's helmet. Not surprisingly, when attacked by five SPECTRE helicopters, Bond manages to destroy all of them using his weapons.

Of all the unusual vehicles used by James Bond in the film series, Little Nellie seemed the most outrageous and improbable. A one-man open helicopter that could soar like an eagle and yet carry a full load of armament, the autogyro seemed like something dreamed up by an insane aerodynamic engineer with too much time on his hands. Still, of all the many Bond travel devices, Little Nellie was the most authentic. It actually existed and flew on real missions in addition to its movie adventures. In 1931, Amelia Earhart, the famous American aviator, flew a similar model, a Pitcairn PCA-2 autogyro, to a then world-altitude record of 18,415 feet. Little Nellie might have been small, but in the aerodynamics field, it had a big reputation.

The easiest way to understand autogyros is to consider them as part of the bigger picture. There are two broad categories of aircraft. The first, fixed-wing aircraft, are airplanes; while the second type, rotary-wing aircraft, are helicopters and autogyros. The word *helicopter* is derived from the Greek words *helix* (spiral) and *pteron* (wing).

Fixed-wing aircraft use an internal-combustion engine and a propeller, or a jet engine, to move the plane and generate lift (a detailed explanation of lift is given in chapter 3). A helicopter is an aircraft that is lifted and propelled by one or two large parallel-to-the-ground rotors (propellers). The movement of air past the rotor causes the lift.

Autogyros have existed for more than eighty years. They have also been called gyrocopters, gyroplanes, and autogiros. They're the first rotary-wing aircraft to fly successfully. By design, they are safer than airplanes and capable of better low-speed flight than planes. They also have the advantage of being capable of vertical takeoff and landing. Unfortunately, autogyros are neither efficient nor fast. Fixed-wing aircraft use less fuel over the same distance. Still, autogyros were invented before helicopters and are fairly easy to assemble from kits. They are, in many ways, the perfect vehicle for spies like James Bond.

An autogyro uses a rotor instead of wings to provide lift. Unlike a helicopter, however, the autogyro rotor is not powered by the engine. It spins due to a force known as *autorotation*. Wind traveling across the rotors causes autorotation, so an autogyro needs an engine to start it moving across the ground. Usually, these engines have been attached to propellers to get the airship moving, but it's possible to use jet engines as well.

Autorotation is caused by wind passing through the rotors of the forward-moving gyrocopter. Once the rotor starts spinning, it generates what is called the relative wind due to the rotor. Since the aircraft's motor causes the vehicle to move forward, there is also a relative wind due to the aircraft. If we add together the force of these two winds, we get what is called the resultant relative wind.

Now, as is the case with airplanes, any wind passing over an airfoil (a wing or rotor) will create both lift and drag. The lift will be perpendicular to the airflow, and the drag will be parallel to the airflow. This fact is true for all airfoils, not just for the rotor in an autogyro. For a more complete description of how lift works, we highly recommend the chapter on "the Vulture" in our book *The Science of Supervillains*.

If we add together the lift and drag forces, they combine to form a resultant force, which is located in front of the axis of rotation of the autogyro. Thus, the resultant relative wind both provides lift and pulls the rotor forward. And thus the autogyro flies.

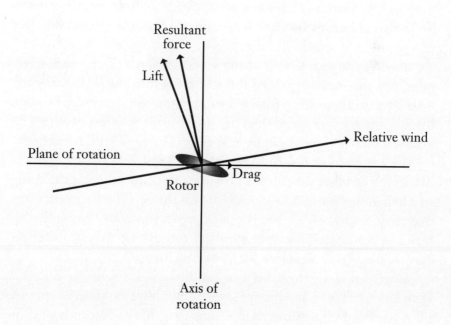

Because of their unique design, autogyros are lifted upward very quickly by their rotors and thus do not require as much area to take off and land as do airplanes. Early autogyros required only about 50 feet of runway to take off and could land in under 20 feet, while airplanes were using hundreds of feet. Later autogyros reduced their need for a runway to less than 15 feet, and eventually, with the help of a second motor for the rotor, were able to take off and land vertically. This unique design feature allowed autogyros to be flown from practically anywhere, needing almost no runway. In fact, in the 1930s and 1940s, autogyros were used to carry mail from post office rooftops in Camden, New Jersey; Philadelphia; Chicago; New Orleans; and Washington, D.C., as well as in other cities in the Northeast.[5]

Autogyros are unique in their ability to fly slowly and not stall. The amount of lift generated by an airfoil (rotor) is proportional to the relative speed at which it moves through the air and to the angle at which it's moving through the air. When the nose of an airplane rises above a certain angle, there is a sharp decrease in the amount of lift made by the wing. This is known as stall and is caused by the air not being able to change direction fast enough to stay attached to the wing. Since the wings on an airplane are fixed, creating more lift means either moving the whole aircraft faster or increasing the wings' angle of attack. In an autogyro, the wings are the rotor and are moving through the air at the speed at which the rotor is spinning, not the speed at which the aircraft is moving. The autogyro does have to be moving forward to maintain the autorotation, but this is a much lower speed than the speed airplanes must maintain to produce lift. Autogyros can fly at speeds as low as 15 miles per hour.

An autogyro can also fly faster than most helicopters. This is due to the fact that the rotor is providing only lift, whereas the rotor in a helicopter is providing both lift and thrust. The helicopter has one major advantage over the autogyro, however: the ability to hover. The forward movement of the autogyro is necessary for its autorotation, and therefore for sustained lift.

When an autogyro stops moving forward, it begins to descend. If an engine fails in an autogyro, the same thing happens as when a pilot tries to fly too slowly. The autogyro slowly descends until it lands. In fact, the procedure for landing an autogyro after engine failure is the same as that for landing an autogyro under ordinary circumstances.

The wing commander Ken Wallis designed Little Nellie for *You Only Live Twice*. Wallis not only built autogyros, he flew them. He held most of the world records for autogyro flying, including the fastest: 111.7 mph. He also flew the longest distance in a straight line: approximately 540 miles.

Autogyros are comparatively safe, fairly easy to build, and extremely easy to fly. They're even used by cowboys in Australia to round up animals. Unlike the jetpack from *Thunderball*, Little Nellie is a real-life James Bond flying machine. It deserves to be used again.

Hydrofoil Boats

After the astonishing worldwide success of *Goldfinger*, Eon Studios rushed into production the next film in the series, *Thunderball*, for a Christmas 1965 opening. The movie freely borrowed from other Bond plots, the villains were not much more menacing, and the crime was slightly less ambitious than in *Goldfinger*: pay SPECTRE a hundred million dollars in diamonds or they will detonate one of two hijacked nuclear warheads in a major U.S. or British city. Audiences didn't seem to care that *Thunderball* was formulaic. The movie grossed over $63 million in its U.S. release and $141 million worldwide. Adjusted to today's market, the worldwide total puts the film at over $800 million in receipts, an all-time record for a Bond movie.

Bond is just one of dozens of agents sent to locate the bomb and end the nuclear blackmail. Mystery has very little to do with the story, as does spying. We all know fairly soon who the villains are and how Bond intends to foil their plans. There's a big battle underwater with lots of neat fighting, and, in a last-second thrill ride, Emilio Largo's boat converts into a hydrofoil, the fastest boat in the world.

Still, this unexpected revelation adds little to the movie. That the final battle between Bond and Largo occurs on a ship traveling at 52 knots per hour (kph) means nothing. The disclosure of the hydrofoil is merely another wonderful gadget for fans. Since it can't outrun radar or planes, the boat offers Largo no last-minute reprieve. The hydrofoil's speed is not an element woven into the plot; it is not significant. There's no surprise attack planned on Miami Beach, no escape to some secret base in South America, nothing. The hydrofoil is there because it looks good. It's science masquerading as entertainment.

Despite the futuristic look of Largo's hydrofoil in *Thunderball*, the boat is nothing new. Hydrofoils have been around for nearly a hundred years. For all of their speed and beauty, they've just never caught on with the American public.

A hydrofoil is a boat that blends aerodynamics and hydrodynamics. The craft looks like a ship that is raised off the water by small, winglike foils. The foils function through Bernoulli's principle. Water flowing over the upper, curved surface of the foil has to move faster than water flowing beneath, so that both streams meet at the back at the same time. The faster-moving fluid on top of the wing creates low pressure above the foil, while the slower moving water beneath creates high pressure. Thus, the foil in motion, just as with an airplane wing, lifts the hull of the ship from the water. With only the foils in the water, the ship's drag is greatly reduced and its speed significantly rises.

Early hydrofoils used U-shaped foils, but over the years, designers have switched to T-shaped foils because they provide better maneuverability.

Hydrofoils were in development long before airplanes. The Frenchman Emmanuel Denis Farcot patented a hydrofoil boat system in 1869. The earliest record of a successful hydrofoil experiment was in 1894, when the Meacham brothers demonstrated their 14-foot hydrofoil speedboat in Chicago.[6] These early attempts to construct working hydrofoils were hampered by the lack of suitable building materials and powerful enough motors to propel the boats.

In 1905, Enrico Forlanini built a successful hydrofoil, using foils constructed like the rungs of a ladder. His boat could reach a speed of over 30 kph. In 1906, William E. Meacham wrote an article about hydrofoils for *Scientific American* that caught the attention of the famous inventor Alexander Graham Bell. After reading the article, Bell began working on hydrofoils with Casey Baldwin, the manager of his estate and laboratory.

During a world tour in 1910 and 1911, Bell and Baldwin met Forlanini in Italy and rode his hydrofoil boat over Lake Maggiore, which connects Switzerland and Italy. Bell's wife commented that the ride was as smooth as flying.[7]

Upon returning to the United States, Bell and Baldwin continued experimenting with hydrofoils, using increasingly powerful engines. After numerous trials, they named their new hydrofoil the HD-4. Using Renault engines, the HD-4 achieved a top water speed of 47 kph and was able to steer well in choppy water. Based on these results, Bell was able to obtain two 350-horsepower engines from the U.S. Navy. On September 9, 1919, using these two motors for power, the HD-4 reached a speed of over 60 kph, a world marine speed record that stood for ten years.[8] The officers at the demonstration, however, felt hydrofoils were too fragile for the navy and refused to fund further research.

Hanns von Schertal was considered the father of the modern hydrofoil. With new foil designs, he overcame propulsion and stability problems for larger ships. In 1940, he was awarded a contract by the German navy to build a 17-ton hydrofoil minelayer. The boat reached an ocean speed of 47 kph, a record that remained unbroken for the next twenty-five years.

It wasn't until 1953 that the commercial value of the hydrofoil was finally recognized. On Lake Maggiore, a 10-ton, 28-passenger von Schertal hydrofoil took just 48 minutes to cross the lake. Normally, the voyage by ferry took 3 hours. In 1956, Carlo Rodriguez, the head of Sicily's largest shipyard, built several hydrofoils using von Schertal's design. These hydrofoils were used to transport seventy-five passengers between Sicily and Italy. In the four years following its introduction to the market, more than a million people traveled by hydrofoil.[9]

In the 1970s and 1980s, there was a renewed interest in hydrofoil boats for use as ferries in the Soviet Union. A number of stream-lined designs were introduced, including the Meteor and the Vokhod boats. The success of these two models has prompted a renewed interest in speed ferries throughout the world. A turbojet ferry transports passengers across the Pearl River delta between Hong Kong and Macau in just over an hour, traveling at an average speed of 60 kph.[10]

In *Thunderball*, Largo's hydrofoil, disguised as a normal ship for most of the film, was just another gadget; it was an interesting but ultimately unimportant part of the film. In other Bond films, the villains and Bond race on water using ordinary speedboats. It's too bad that Q Division ignores hydrofoils; used properly, they could be a fast and deadly part of the Bond arsenal.

The Stealth Boat

In *Tomorrow Never Dies*, the eighteenth James Bond movie, our hero fights the media mogul Elliot Carver, who plans to bring the world to the brink of nuclear Armageddon to help expand his global news empire. Carver, who owns worldwide TV stations, radio stations, and newspapers, also controls a stealth battleship loaded with torpe-does and rockets. The boat, invisible to radar, sinks a British destroyer off the coast of China, and its crew steals a nuclear missile from the sunken ship's hold. Carver plans to blow up Beijing with the British bomb. A Chinese general, who is part of the plan, will negotiate peace and grant Carver exclusive media rights to China for the next century. It's a complicated plot with room for human error, but by 1997, no one expected the Bond films to be the least bit believable. The story line was merely an excuse for action and excitement.

As is the case in several previous Bond adventures, the plot hinges on the existence of a secret ship. In *You Only Live Twice*, Blofeld uses his own spaceship, launched from the inside of an extinct volcano, to capture U.S. and Soviet space capsules. In *The Spy Who Loved Me*, Karl Stromberg uses a giant oil tanker, the *Liparus*, to capture nuclear submarines so he can use their missiles in another attempt to create a global holocaust. In *Tomorrow Never Dies*, Carver relies on an extremely fast stealth warship in his attempt to trick the British and Chinese into war. None of these crafts seems the least bit believable. Yet, the most incredible of them all, Carver's stealth ship, is closer to reality than most people realize.

In World War I, the world came to realize how vulnerable modern ships were to sneak attacks. In the North Sea, German submarines easily torpedoed many Allied ships. In World War II, submarines and their torpedoes took a heavy toll on shipping, and aircraft bombing was also very effective. Large, slow-moving ships made easy targets.

To make them more resistant to such attacks, modern warships are built with thick steel hulls. However, weapons have improved faster than armor. Rocket weapons are extremely sophisticated and carry a devastating payload. A single missile can disable a ship. A dozen missiles can destroy one. On October 12, 2000, the terrorist attack on the USS *Cole* proved that even large ships are vulnerable to suicide bombers. *Cole* was attacked in a friendly harbor in Yemen. In hostile waters, with dozens of small ships in the area, the missile cruiser would have been in even greater peril. This is why most battleships are useless in land-sea encounters.

Modern stealth technology is used to prevent the detection and identification of vessels. Submarines have made use of stealth techniques for decades. Recent advances in battle planes have focused on stealth technology. Now, stealth technology is being developed for boats.

The biggest challenge in the world of espionage is keeping things hidden. Radar is one of the most effective tools for finding things. The basic concept of radar is simple. A radar dish emits a signal with a known spectrum. When that signal strikes an object, it is reflected back to the transmitter. The delay time between when the signal is transmitted and when it is received back is used to calculate the distance of the object from the radar dish. The signal phase can be analyzed to discover if the object being traced is moving and how fast it is going. Constructing a boat that is near-invisible to radar is the goal of modern stealth technology.

Because radar works by receiving the reflection of a signal sent out, there are two methods of possibly fooling radar. One method is to design a ship so that any radar waves striking it reflect off its surface in a totally different direction. (Most radar uses radio waves, a form of electromagnetic energy that moves at the speed of light.) A great deal of stealth technology involves designing structures with sharp angles and flat surfaces that reflect radar waves in different directions.

The other method of fooling radar is to construct structures made from material that absorbs radar waves so that they do not bounce off the ship at all. This involves covering a ship with radar-absorbent material (RAM). When a radar wave hits an object, the wave passes some of its energy to the material, exciting its electrons. In a good conductor, such as a metal antenna, the radar wave doesn't lose much energy, because the electrons move easily and thus reflect back to the source without much reduction. RAM, however, is a poor conductor, and the electrons in it move slowly. The electrons resist moving, so the radio wave loses much of its energy, which is then converted into heat. Thus, the signal reflected off the RAM is greatly reduced.

What exactly is RAM? It's evidently a composite material, a combination of various lightweight substances, specifically designed to absorb radio energy. That's about as specific as anyone will get when discussing such stuff. Some secrets are not open to public scrutiny, leaving us to state that it works, but we're not sure what it is.

The U.S. Navy has been working on a stealth ship for the past decade called the Littoral Combat Ship (LCS). The navy hopes to have the LCS in the water by 2007. But if there's a delay, we can always turn to the Norwegians and buy their P-960 KNM *Skjold*.

The *Skjold*, a hovercraft, is the world's first stealth attack ship. It is approximately 155 feet long, 44 feet wide, and 40 feet high. Riding on a cushion of air, it can reach speeds of up to 55 kph. It carries a 76 mm Oto Melara Super Rapid gun and a full array of anti-ship missiles.

Instead of covering the exterior of the ship with panels of RAM, the boat itself is made of RAM, which greatly reduces the vessel's weight. The ship is designed with a small number of reflective panels and no right-angled corners.

The secondary structure of the ship is flush with the boat's exterior. All doors and hatches are flush with the walls and have the same radar-absorbing characteristics as the walls. Radar-absorbing covers fit over deck material such as mooring pieces.

The *Skjold* was commissioned on April 17, 1999, approximately two years after James Bond's encounter with a stealth ship in *Tomorrow Never Dies*. Amazingly, both Carver's unique vessel and the *Skjold* cruise at a top speed of 55 kph. Coincidence? In the world of James Bond, superspy, is there ever such a thing as coincidence? We'll leave it to you to decide.

8

Getting Farther Away
from It All

Outer-Space Shenanigans

Allow me to introduce you to the airlock chamber. Observe,
Mr. Bond, your route from this world to the next. And the
treacherous Dr. Goodhead; your desire to become America's
first woman in space will shortly be fulfilled.

—HUGO DRAX, *MOONRAKER*

One of the most appealing features of the James Bond movies is
their exotic locations. Unlike most spies, Bond doesn't spend his
time in gritty, dimly lit warehouses or grimy offices in the middle of
London, Copenhagen, or Langley Air Force Base. Instead, as befitting
his role as a spy who handles only the most dangerous missions
for his government, and often the nations of the world, battling
some megalomaniac villain, 007 travels to the most remote and
unusual locations on the globe. His adventures have taken place just
about everywhere on Earth. And sometimes, even beyond.

Toppling Dr. No

In the first James Bond movie, released in 1962, 007 finds himself
involved in the race for outer space. Bond is in London, playing
with cards and women, when he's summoned by his boss, M, to
headquarters. M tells Bond that 006 had been investigating some

sort of massive interference with Cape Canaveral rockets and that this interference is coming from Jamaica. M asks Bond if he knows what *toppling* means, and of course Bond has a ready answer: "throwing gyroscopic controls of the rockets off balance using a radio beam." Apparently, the rockets toppled off balance and ended up in the Brazilian jungle. Now, Britain and the United States are afraid to send rockets around the moon.

Gyroscopes basically use a fast-moving ring inside a gimballed structure. A spinning gyroscope will continue to maintain a certain orientation, the direction in which it is pointing, no matter how you move the frame. Thus, the spinning gyroscope will also serve as a three-dimensional compass.[1] At the start of a trip, the axis of a gyrocompass is pointed north using a magnetic compass as a reference. A motor inside the gyrocompass keeps the gyroscope spinning, so the gyrocompass continues to point north, thus serving as a fixed point of reference. If a gyroscope becomes imbalanced, however, as in *Dr. No*, the navigation system for a rocket can get screwed up.

The first flaw in the plot comes when we try to determine how a radio beam, as used by Dr. No, can screw up the gyroscope. If the beam interferes with radio signals that provide directions to the gyroscope, then the scope would receive incorrect navigational coordinates and could possibly land the rocket in the Brazilian jungle. But we're never told that the radio beam is messing with a bunch of signals. We're told that the beam has set the gyroscope off balance. It's a bit of a stretch.

After further adventures, Bond goes to take a look at Crab Key, an island from where the radio signals could have been beamed to the missiles. The locals say that a Chinese man owns the island. The island has a bauxite mine and is equipped with a low-can radar setup, meaning the radar is very close to the surface and can detect incoming ships. Bond and his guide make it to the island, where they meet Honey Ryder. She's a beautiful young woman who scours the island for large seashells to sell to tourists. It's a bad night for shell hunting when Dr. No's henchmen track down Bond, his guide, and Ryder using a jeep armed with a flamethrower disguised as a dragon. The superstitious guide is burned to death, and Honey and 007 are taken prisoner.

At Dr. No's headquarters everyone is wearing contamination suits. Bond and Ryder are hosed down, washed, and scrubbed with soap. Geiger counters show that their bodies are radioactive at a level of 72.8. They are stripped of clothes, then put through a shower of high-pressure jets, one after another, with more soap. Finally, the Geiger counters show that their bodies are no longer radioactive.

Just what does the Geiger counter measure, and if Bond and Ryder are radioactive at a level of 72.8, are they supposed to be dead, fiercely ill, or contaminated, or is it possible that soap and high-pressure sprays can safely decontaminate them with no ill side effects?

First, let's define some terms. Electromagnetic radiation is a form of energy that moves through space, and it includes X-rays, light, and Dr. No's illusory radio waves. Gamma rays are extremely dangerous. Low-energy radio waves pass through our bodies without causing serious problems. Ionizing radiation makes an atom charged. It occurs when an unstable atom's nucleus decays and releases its energy as ionizing radiation: alpha, beta, or gamma.

Geiger counters are ionization counters. But what does a count of 72.8 mean? Radiation is generally measured in 0.001 sieverts, or 1 millisievert (mSv). A chest X-ray supplies 0.3 mSv. The recommended occupational limit is 20 mSv; more than that, and you shouldn't go to work. At Hiroshima and Nagasaki, after the bombs fell, the average dosage was 200 mSv. Acute symptoms and near-certain death occur with a dosage of 4,500 mSv, but frankly, you wouldn't want to be exposed to 200 mSv. Is it possible that Bond and Ryder get a dose of 72.8 mSv—that is, nearly four times the recommended annual occupational limit or approximately a third of the dosage received after the bombs dropped at Hiroshima and Nagasaki?

A sievert is similar to a rem, with 1 sievert equal to 100 rem and 1 mSv equal to 0.1 rem.[2] In short, rem stands for "roentgen equivalent man." A dose of sievert is similar to a dose in $Gy \times Q$. One gray (Gy) is equal to 100 rads, and 1 rad is equal to 0.01 Gy. Q is a factor that denotes the type of radiation and its level of damage. 800 rems means certain death and is caused by high levels of gamma radiation.

So let's juggle the numbers, something we always enjoy doing:

$$1 \text{ mSv} = 0.1 \text{ rem}$$
$$72.8 \text{ mSv} = (72.8 \times 0.1) \text{ Rem} = 7.28 \text{ rem}$$

Our answer, then, is that 72.8 mSv is not going to kill Bond and Ryder. But if they're getting a third of the dosage received after the bombs dropped at Hiroshima and Nagasaki, they have reason for concern.

In the main control room, Bond encounters remote manipulators that handle bizarre computer equipment. Giant tape drives are whirling. Oscilloscopes are everywhere, showing us electric signal patterns. The reactor is on. Dr. No is using power from the nuclear reactor to generate the giant signal that will throw the rocket gyroscopes totally off balance, hurtling the rockets into the jungle. Here is a second flaw in the plot: each rocket probably has preprogrammed navigation, and while the rocket is lifting into the atmosphere, the noise would jam any transmission of information to the rocket. Once the rocket is cruising, it might be possible to recalibrate it and send it a signal for a new direction. But during liftoff, the rocket is most likely self-regulating, and so it makes no sense to try and disrupt some sort of navigational signal to the rocket.

Of course, since we've already been asked to believe that fishermen in Jamaica think a jeep with a flamethrower and some decorations is a real dragon, it's not a stretch to assume that Dr. No's radio device will knock the rockets off course. As in most Bond movies, the basic notion is to make the science look and sound *believable* without ever checking whether it's even remotely *possible*. Bond kills Dr. No, escapes with Ryder, and the villain's hideaway blows up. It's the first Bond movie, and 007's first encounter with rockets and spaceships, but it's definitely not his last.

The *You Only Live Twice* Gobbling Spaceship

Starring Sean Connery, 1967's *You Only Live Twice* features a space-ship that swallows other ships and hence kidnaps astronauts. The movie begins with an astronaut floating in outer space while con-nected to a rocket ship, *Jupiter 16*, via a life belt. An unidentified object appears on the command center scope in Hawaii. The object is speeding toward the ship to which the astronaut is attached. As the men in *Jupiter 16* see the approaching object, they realize it's another spaceship. The front of the unidentified ship opens like the mouth of a giant shark. And like a giant shark, the spaceship gobbles *Jupiter 16*, snapping the lifeline to the astronaut, who floats away.

The United States assumes that the Soviet Union is responsible for launching the gobbling spaceship. If another U.S. spaceship is eaten, the U.S. government will consider it an act of war.

In the meantime, at bad guy Ernst Stavro Blofeld's head-quarters, the gobbling spaceship opens its mouth and some grap-ples lift a Soviet rocket ship from the maw. It's all an evil plan by the SPECTRE chief Blofeld: the mainland Chinese will give him $100 million in gold bullion if he can create a war between the United States and the Soviet Union, a nuclear war from which the Chinese expect to emerge victorious. (To see how ridiculous this plan is, read chapter 5.)

Later, in an effort to infiltrate the evilmongers, Bond is dolled up to look like a Japanese man—well, sort of. His hair is dyed black and cut straight across the forehead. His eyebrows are plucked. He still looks like a Brit, but we pretend that he looks Japanese. He trains to become a Ninja, and he fakes marriage to a traditional Japanese woman.

At the end, Bond and the beautiful (of course) Japanese agent Kissy Suzuki team with Ninja warriors to raid Blofeld's lair, which is located in an inactive volcano.

A few questions immediately come to mind when trying to determine if a gobbling spaceship can exist. The first question is composed of several parts: can the gobbling spaceship maneuver with sufficient precision to navigate toward another ship in outer space, pause exactly in front of the other ship, and then eat the other ship without a major explosion taking place?

Our first reaction is that there's no way a spaceship can navigate with that amount of precision in outer space. For one thing, space-ships travel extremely fast, and coming up on another ship—bam, presto, without notice—implies that the gobbler is moving as quickly as, if not more quickly than, "ordinary" spaceships. If it's moving that fast, how does it suddenly pause right in front of the other ship and gobble it up?

Robert Hutchings Goddard, for whom NASA's Goddard Spaceflight Center is named, was a rocket pioneer. In 1915, he designed experiments proving that rockets could operate in vacu-ums, and therefore that they could operate in outer space. In 1921, he began working on fuel injection and ignition, as well as on engine cooling for rocket systems. Then in 1926 he launched a liquid-propelled rocket, which flew at 60 mph.[3] It is possible that a gobbler at 60 mph could pause directly in front of another ship, but then the gobbler couldn't simultaneously be zooming so quickly that it remains unnoticed until the last second. (Unless, of course, the gobbler is cloaked as in *Star Trek*, but *You Only Live Twice* does-n't suggest that notion at all!)

In 1945, Goddard refined his liquid-fuel rockets until they were speeding at 300 mph. If the gobbler reached 300 mph, it's vaguely (and we use the term *vaguely* to mean highly unlikely) conceivable that another ship wouldn't notice it until the gobbler opened its maw a foot away.

But we still don't have a logical explanation for how the gobbler can move quickly yet halt instantly and eat another spaceship. So let's move forward in time, analyzing rockets and their speeds and the science behind space travel.

Rocket motors burn fuel, producing a lot of gas. The gas bellows out of the rocket at one end, which propels the rocket forward in the other direction. The force of travel is derived from the momentum of all the gas molecules that are bellowing out of the rocket. Isaac Newton's third law of motion tells us that the rocket's speed in the other direction matches the momentum of these gas molecules.

When a rocket is moving at between 5,000 and 10,000 mph, most of the gas molecules are expelled from the bottom of the rocket. To escape Earth's gravity and go into an outer space orbit, a rocket must reach a velocity of 17,400 mph.[4] So already we know that our gobbler must be going at least 17,400 mph. We might be ready to believe that the gobbler wouldn't be noticed at this speed, but then how do we explain that we are able to track such rockets from Earth today? Hence, even at this speed, the gobbler still doesn't make sense.

We note that to attain sufficient speed to enter orbit, a rocket has several stages. It has lighter rockets on top of heavier ones. Each stage launches as the entire rocket goes upward, and the velocity increases cumulatively. Many modern rockets use booster rockets, which supply more fuel and power, then fall off the main rocket assembly into Earth's oceans. GPS satellites, for example, use booster rockets.

A few other methods exist for propelling rockets through space. (First, of course, the rocket has to make it into space.) Electrothermal rocket engines use electrically charged or ionized gas to heat propellants. Electrostatic rocket engines ionize the propellants and then zap them through electric fields. Electrodynamic engines generate electrically charged or ionized gas and then zap it through electric or magnetic fields. Thermonuclear engines, which are still being tested, use nuclear reactors to heat propellants.[5] It is thought (by some scientists and many science-fiction writers) that someday we'll use matter-antimatter systems and nuclear fusion engines to propel our spaceships, but these notions are a long way off. Even so, we wonder if the *You Only Live Twice* gobbling spaceship used a futuristic method of travel, as real science doesn't seem to explain how the gobbler zooms so quickly we don't notice it, only to stop mere inches from another ship.

Speaking of inches, is it possible for a spacecraft to navigate carefully enough through space to stop mere inches from another ship? Astronomers map outer space with coordinates beyond our Earth's three-dimensional x, y, and z grid. They plot space with coordinates such as azimuth, the direction from the north; right ascension, which is like longitude; elevation, the angle above the horizon; distance; and time.

To navigate through space, one does not take a left on Main Street and a right on North Street and go west on River Road. Rather, one starts at the center of Earth, with the z axis going up and the x and y axes measured from the equator. Earth's rotation must be taken into account. A computer on the spaceship analyzes the position of the ship as it moves through space. The computer changes the position based on the coordinates noted earlier, as well as on x, y, z, and Earth's rotation.

In addition, most of the navigation work occurs before the spacecraft is launched. Keeping in mind that objects in space are constantly moving, navigation up there isn't as simple as on Earth, where buildings generally do not move across town as you're driving down the road. If that doesn't complicate things enough, the gravity exerted by planets and the sun bends the flights of spaceships. Scientists plan a spaceship's course, taking into account the gravity of planets, moons, and the sun. Basically, once a ship is launched, it's tough to make it turn around and travel in the opposite direction.

Once the spaceship is in orbit, however, some minor adjustments *can* be made to the flight course. But these adjustments generally correct a course to realign a craft along the predetermined path. If a correction is needed in the flight path, the spaceship fires small rockets that slightly change the ship's direction. On the *Deep Space 1* craft, which we'll describe in a moment, autonavigation can handle these small corrections.

For now, it's important to note that if the path of a spaceship is altered, the ship will continue to move off course, until, ultimately, it's far away from the original destination. We can't ignore Newton's second law of motion, which states that a force (our rockets) acting on an object (in this case, our spaceship) gives it an acceleration in the direction of the force, with a magnitude inversely proportional to the mass of the object. In other words, big mistakes can't be easily fixed in outer space navigation. A spaceship doesn't have the fuel to revise course, go in the opposite direction, then shift course again and zoom back to its original path. And remember, destinations and navigational objects such as planets, moons, and so on are constantly moving in outer space.

It seems unlikely that the gobbler can navigate carefully enough through space to stop mere inches from another ship that is also in constant motion, gobble it, then back up, and resume its original course.

Deep Space 1 was launched in October 1998 and tested twelve advanced space technologies. In 1999, it flew past the asteroid Braille, coming within 16 miles to collect scientific data.[6] This maneuver is considered extraordinary. Still, even *Deep Space 1* could not have eaten Braille from 16 miles away!

As for private spacecraft, in June 2004, *SpaceShipOne* became the first civilian vehicle to fly into outer space.[7] A mother craft climbed to 46,000 feet (or 7.58 miles), then released *SpaceShipOne*, which reached an altitude of 62 miles, the "official edge of space," as reported by *New Scientist*.[8] The official edge is defined as a suborbital flight, and *SpaceShipOne*'s lead designer, Burt Rutan of the Scaled Composites of Mojave Company, acknowledges that reaching orbit will be a much more difficult task.[9] The ultimate hope is for a new version of the private craft to carry from six to ten tourists at a time, all of them adventurers with a lot of money to burn.

The *You Only Live Twice* gobbling spaceship must travel much faster than the ship it eats, and it must possess extraordinary navigation. How could the gobbling spaceship maneuver so cleverly and precisely in outer space to find speeding rockets, get exactly on the right track for interception, and gobble them up without a major explosion? The answer is simple: it couldn't.

The Moonraker Space Station

In 1976, Cubby Broccoli took complete control of the James Bond film franchise, buying out his partner, Harry Saltzman. At the time, Broccoli announced that he planned to release one Bond film per year. *The Spy Who Loved Me* was already scheduled for release in 1977. It was to be followed in 1978 by *Moonraker* and *For Your Eyes Only* in 1979.

The best laid plans of mice and producers oft go wrong, however, and during the end credits of *The Spy Who Loved Me*, the tagline read, "James Bond will return next in *For Your Eyes Only*." The reason for the schedule shift seemed obvious. *Moonraker* was based on the U.S. space shuttle mission, which was bogged down in delays, and the plot of *Moonraker*, with its madman wanting to rule the globe, was very similar to *The Spy Who Loved Me*. Two Bond movies in a row featuring megalomaniacs implied a certain staleness in theme.

Then, in 1977, one movie changed cinema history. *Star Wars* opened and broke all sorts of box-office records. The world was suddenly fascinated by outer space, light sabers, laser guns, and spaceships. *Moonraker* moved back on schedule as the next Bond film to follow *The Spy Who Loved Me*.

Moonraker premiered in 1979 and was the eleventh James Bond film. Of all the Bond films, *Moonraker* is perhaps the most debated among fans. A large number of Bond fanatics who started watching the movies in the late 1970s and early 1980s consider *Moonraker* one of the best Roger Moore adventures and one of the better entries in the Bond series. The film featured plenty of Bond action, laser battles in outer space, tongue-in-cheek humor, and a complex plot about poisoning the world. It even featured the return of Richard Kiel as the steel-jawed assassin Jaws.

An equal number of fans thought the film was the worst entry in the Bond series, rivaled only by the equally inept *Diamonds Are Forever*. What *Moonraker* lacked was any sense of reality or logic. The story jumped from place to place across the globe, based more on Broccoli's fondness for certain locations (such as the Brazilian rain forest) than on any requirements of the story. The plot made little sense and bent and sometimes broke the basic rules of science in hopes of telling an exciting story. The final scene, which took place on a "radar invisible" space station, was particularly outrageous.[10]

We leave it to our readers to reach their own conclusions. All we will state is that *Moonraker* featured some of the most advanced scientific concepts ever shown in the James Bond series.

Unfortunately, in almost every case the script used these concepts in the worst possible manner. *Moonraker* wasn't merely bad science; it was bad science fiction.

When a jumbo jet carrying a space shuttle lent to the British by the United States crashes while crossing the ocean, James Bond is sent to investigate. In the film, the space shuttle, dubbed *Moonraker*, is capable of being blasted into space on the back of a three-stage rocket, detaching from the rocket and flying about in space, then making a soft landing on Earth. Shown on a movie screen two years before the United States ever launched a space shuttle, the *Moonraker* project is a marvel of motion picture prediction. The film got many of the shuttle details right, except for the concept that the shuttle would be used for transportation, as some sort of bus in outer space, rather than for carrying cargo.

When Bond studies the crash site, he discovers that there's no debris present from the shuttle. It doesn't take Bond long to realize that someone stole the spacecraft and crashed the plane to cover up the theft. Soon our hero is investigating the employees of Drax Industries in California, the company that built the shuttle. Hugo Drax, the wealthy owner of the corporation, assures Bond he's wasting his time, but we suspect otherwise—as does Bond.

At the company compound, Bond is strapped into a high-gravity simulator booth by Dr. Holly Goodhead, who appears to be a female astronaut and part-time supermodel. The booth, a standard feature in numerous spaceflight movies, is attached to a gigantic rotor that spins it around faster and faster. The high speed increases the gravitational pressure on anyone sitting in the booth. A g is the force of 1 Earth gravity; at 7 g a person weighs seven times his or her normal weight. G pressure is dangerous because as the g increases, it takes more effort for a heart to pump blood through a person's body.

Dr. Goodhead tells Bond that 3 g is the pressure felt on a spaceship takeoff. Most people pass out at 7 g. The machine is capable of going up to 20 g, a lethal speed. Needless to say, Drax sends his assistant to kill Bond by pushing the machine up to 15 g. A circuit blows, and the machine stops before our hero is badly injured.

After some initial distrust, Bond and Dr. Goodhead team up to battle Drax. The winding trail takes them to Europe and then to South America. It's there that they discover the villain's secret base in the heart of the Amazon rain forest. Drax plans to kill all humankind with a superpoison (see chapter 11 for details about Drax's orchids). Once Earth is cleansed of ordinary people, Drax plans to repopulate the planet with a superior race of perfect humans who will be safely protected on his space station during the release of the superpoison.

Drax reveals to Bond that he stole the *Moonraker* that his company built for the United States because one of his private spaceships was damaged. He then departs for his space station, leaving Bond and Dr. Goodhead to be charred to ashes by a rocket's exhaust blast. After all, why merely shoot the hero when you can leave him unguarded in a pit for several minutes? Such is the thinking of criminal geniuses.

After much action, Bond and Dr. Goodhead escape being fried and sneak onto *Moonraker 6*. The space shuttle is preprogrammed, so all Bond and Dr. Goodhead need do is watch as the ship rendezvouses with a giant space station that's invisible to radar. (We discuss stealth technology in chapter 7.) While the space station can't be seen by radar, the six *Moonraker* spaceships docked at it can be, which leaves a rather large hole in the plot. But, after the Death Orchids and the scene in the rocket exhaust chamber, logic has already leaped off a cliff, so no one notices.

All of the *Moonrakers* dock on the space station, and their crews of genetically perfect humans (mostly blonds with blue eyes, though there is a nod to some racial diversity) gather on the main deck. Bond and Dr. Goodhead find the Orbital Communication Headquarters on level 10 of the station and locate Drax's radar jamming system, which they immediately shut off.

Suddenly, the space station appears on NASA radar screens in the United States (which are aimed at outer space, evidently scanning the skies for the flying saucers due on Independence Day). Of course, once the space station is spotted, the United States immediately launches a space shuttle filled with marines in space suits to investigate.

Meanwhile, Drax plans to bombard Earth with fifty poison globes, each capable of killing 100 million people. In a scene almost too cheesy to discuss, Bond stops Drax from launching any more globes into the atmosphere. Jaws undergoes a miraculous conversion brought about by love and decides to help the good guys. Gravity rises and falls on the space station. Drax's guards, in space suits, leave the space fortress to engage in a laser gun battle with the U.S. space marines. Drax is killed by Bond, who launches the billionaire into his own orbit around Earth. The space station catches fire in outer space. All the blond genetic superhumans run away or somehow disappear. At the end, Bond and Dr. Goodhead use a spare *Moonraker* spaceship to blow up the three disease-carrying globes. Logic jumps off a cliff, and the movie mercifully comes to an end, with the usual salacious double-entendre made by Q.

Believe it or not, buried in all this insanity are some nuggets of real science. In 1979, when *Moonraker* was released, the United States and the Soviet Union had already put decades of hard work into outer space exploration projects. In the 1970s, the two superpowers attempted, and in some cases succeeded, in launching various space stations. So, as with many Bond movies, *Moonraker* was right on target in regard to what film audiences wanted to see. *Star Wars* was a huge success, and showing good and bad guys shooting laser guns at one another in outer space became de rigueur.

In reality circa 2004, robots have descended into the electromagnetic fields of Jupiter; and Pioneers, Pathfinders, Mariners, Vikings, and Voyagers have probed, retrieved samples, snapped and transmitted images, and analyzed data from all over the solar system. Humans have walked on the moon and, indirectly via probes, on two other worlds beyond Earth: Venus and Mars.

The 1950s were filled with articles and TV specials discussing space stations orbiting Earth, where large crews of men and women would work for months and sometimes years, solving scientific problems that could not be addressed on Earth. In most scenarios, the space station resembled a huge bicycle wheel, connected by a number of spokes and with a hub at the center. The wheel shape was best, because if it was kept spinning at a constant rotation, centrifugal force would create artificial gravity on the space station. As envisioned by scientists, the space station would enable humans to establish a foothold beyond Earth's gravity, making it easy to send manned expeditions to the moon, Mars, and beyond.

It didn't happen at that time. The cost of building a huge space station in outer space was beyond government budget outlays. Furthermore, what was originally seen as a nonpolitical issue soon became a worldwide challenge: the space race. Stops in space were ignored in the quest for the big prize: landing on the moon by a direct flight from Earth.

In October 1957, the Soviet Union launched *Sputnik 1*, which marked the beginning of the space age and the serious advent of the space race.[11] Newspapers and magazines lamented that American scientists had let the country down, and a new incentive to get students to study math and science was established. Everyone knew that the United States had to be first in the space race. To think otherwise would be un-American.

Soon after *Sputnik*, mice, monkeys, apes, and dogs were launched into outer space. (Sadly, the first dog to enter outer space was a stray named Laika, who died in orbit.)[12]

In 1958, in an effort to keep pace with the Soviet Union, the United States established NASA. By 1959, the Soviet Union was crashing rockets onto the surface of the moon and taking photographs of its far side. And in 1961, the Russian cosmonaut Yuri Gagarin flew into outer space and orbited Earth. Also in 1961, President John F. Kennedy declared that the United States would send a man to the moon before the decade ended.

Several years later, in 1966, Russia landed a robotic space probe on the moon. Both the United States and Russia made regular, manned flights into Earth's orbit, analyzing and refining methods for docking space shuttles and maneuvering rockets in outer space.

Despite years of research into docking and maneuvering spacecrafts, when *Apollo 1* was set to launch in 1967, a fire exploded inside the rocket and killed the crew. A mere four months later, another disaster struck when the Russian *Soyuz 1* crashed, killing its pilot, Vladimir Komarov.

In October 1968, *Apollo 7* was scheduled to test its command and service modules in Earth's orbit. This was almost a full decade after the Soviet Union was crashing rockets on the moon's surface. Also in 1968, the Russians sent a *Soyuz* rocket to the far side of the moon.

At this point, the Russians were ready to send a manned rocket around the moon. NASA scheduled *Apollo 8* to orbit the moon as well. This would be the first time that humans left Earth's orbit and ventured into the vast realm of outer space.

Shortly thereafter, in 1969 *Apollo 11* landed in the Sea of Tranquility on the moon. Then in 1971, the United States put its *Skylab* space station in Earth's orbit.

In 1975, the United States and the Soviet Union linked up their *Apollo* and *Soyuz* spaceships to display world solidarity; after that, manned exploration of outer space ceased until 1981.

In 1981, the United States launched the first space shuttle, which enabled astronauts to take short and frequent trips into Earth's orbit and land back on Earth.[13] This procedure was similar to *Moonraker*'s strategy of launching, orbiting, and then reentering the atmosphere and landing on Earth. But as we'll see soon, *Moonraker*'s technical sophistication far exceeded anything in reality.

Let's rewind to *Skylab 1*, America's first space station. Built from Apollo technology, *Skylab* was expected to launch into orbit from the first two stages of a Saturn V rocket. The lower section of *Skylab* contained a communal area, a toilet, a kitchen, an exercise area, and individual bedrooms. The upper section held a scientific laboratory. Tiny by *Moonraker* comparison, *Skylab 1* weighed 75 tons.

In May 1973, the United States launched the unmanned *Skylab 1*, and disaster struck: a thin sun shield opened early and was ripped off, and with it a solar panel also fell off the spacecraft. When the solar panel fell off, a second panel was damaged. *Skylab 1* continued into Earth's orbit, but without sun shields the temperature was fiercely hot inside the craft, and NASA had to delay its first manned mission, which had been scheduled for the next day.

A week and a half later, a small crew flew to *Skylab 1* to repair the damage. Led by Charles Conrad, an astronaut who had flown on

Apollo 12 and been on the moon, the repair team fixed *Skylab 1* and remained on board for twenty-eight days.

In July 1973, another crew arrived at *Skylab 1*. Led by Alan Bean, this second team remained on *Skylab* for fifty-nine days, during which they further repaired the sun shields. In addition, they conducted research about space medicine and astronomy.

In November 1973, a third crew landed at *Skylab 1*. They stayed for eighty-four days, and the main purpose of their visit was to see how space life affected the human body. This was the last crew to visit *Skylab 1*.

In July 1979, *Skylab 1* crashed into the Indian Ocean. This was the condition of the U.S. space station program in 1979, when *Moonraker* was released.

Clearly, reality in 1979 was much different from *Moonraker*. The few attempts at creating space stations relied on spaceships already in orbit. Living space was at a premium. Those big wheels imagined in the 1950s never made it off the drawing board. Entire populations could not live on space stations.

In fact, reality now is nothing like *Moonraker*, either. Let's review some post-*Moonraker* space station progress.

In 1986, the *Challenger* space shuttle exploded during launch, and the entire crew was killed. Prior to that event, the Russians launched the space station *Salyut 7* in 1982 and sent cosmonauts there for increasingly long periods of time. Svetlana Savitskaya stayed on *Salyut 7* that year and was the first woman ever to walk in space. But then in 1984 technical problems hit *Salyut 7*, and in 1985 Vladimir Dzhanibekov and Victor Savinykh flew on the Russian *Soyuz T13* to repair the *Salyut 7*. The following year, 1986, a final crew visited *Salyut 7*. In 1991, *Salyut 7* crashed to Earth.

Moonraker was a superadvanced (and superstable) version of a space station that reminded us of the Russian *Mir* space station. *Mir* meant "a community living in harmony and peace." It was launched in 1986, and its first two crew members were Leonid Kizim and Vladimir Solovyev. From 1987 to 1989, *Mir* was inhabited almost continuously by various crew members. Two people, Musa Manarov and Vladimir Titov, stayed aboard *Mir* for one full year. One

Afghan cosmonaut and one Frenchman visited *Mir* during this time.

Starting in 1989, crews of two people stayed on *Mir* for six months each. A third crew member would arrive with each two-person crew but then leave with the previous team. Japanese and British citizens stayed on *Mir* for one-week visits.

Mir was far more stable than the *Challenger* space shuttle, but still, *Mir* was nothing like *Moonraker* in terms of sophistication and technical stability. In 1993, the United States announced plans to send ten space shuttle flights to *Mir*, but in the late 1990s, *Mir* was hit by a series of technical problems.

By 1999, plans were under way to turn the failing *Mir* into a space hotel, but even at $20 million for a one-week stay, it was hard for the new hotel to earn enough money to cover the repair bills. In 2001, *Mir* retired from its role as the Grand Hotel of the Universe.

Perhaps it was President Ronald Reagan who saw the technical beauty of *Moonraker* more than anyone else. In 1984, Reagan announced plans to build an international space station to be used by astronauts from all over the world. In 1993, Russia joined the effort. In late 1998, Russia launched the first module of the International Space Station, and then in 2000 Russia launched a second module. The United States, Russia, Japan, eleven nations in Europe, Canada, and Brazil all have contributed to the development of the International Space Station.

When completed, the International Space Station is supposed to be four times larger than *Mir*, and it will weigh more than a million pounds. Its near-acre of solar panels will provide electric power to six onboard research laboratories.

As of this writing, the International Space Station is nowhere near completion. According to NASA, the technical configuration may be finished by 2010.[14] When completed, the station will have a crew of seven people, maximum—not hundreds of blond superbeings.

As for whether *Moonraker* can launch, orbit, and then reenter the atmosphere and land on Earth, it's possible, but it would surely be detected on Earth. In reality, when a space shuttle returns to Earth it encounters enormous air resistance. Space shuttles generally reenter from altitudes of approximately 250 miles, and they reenter our atmosphere traveling at approximately 17,000 mph. At 80 miles high, friction heats the craft to 3,000 degrees Fahrenheit. At approximately 62 miles above the Earth, a spaceship reenters Earth's atmosphere. The outside of the shuttle's wings and nose has carbon tiles, and the inside has silicon-compound tiles, both designed to help the shuttle reenter the atmosphere without exploding.

To return to and land on Earth, a space vehicle must travel during a specific period of time. This period is based on predefined criteria, such as sufficient daylight for scientists to track the return, weather conditions, propellants evaporating from the tanks, reentry path, and a complex series of navigation maneuvers that near-guarantees that the vehicle enters the atmosphere at the exact point and time when its flight path can reach the landing site.

So while *Moonraker* can reenter the atmosphere, it would be impossible for it to do so without somebody noticing it.

Even if *Moonraker* could have existed in 1979 (highly unlikely, as we've seen), we wonder how the young couples secretly prepared for their life in space. Where did they train to be lifelong astronauts, and who trained them?

Currently, to be a mission specialist on board a spacecraft, a U.S. citizen must have skills and qualifications that are almost the same as those required for astronaut pilots. Minimum requirements include:[15]

- A bachelor's degree in engineering, biology, physical science, or mathematics
- Three years of technical experience
- 1,000 hours as a jet pilot
- Successful completion of the NASA Class I space physical examination
- One to two years of specialized NASA training, including advanced classes in shuttle systems, orbital dynamics, astronomy, physics, materials processing, mathematics, geology, meteorology, navigation, and oceanography

- Military water survival training, including advanced scuba training
- The ability to swim 75 meters in a flight suit and tennis shoes
- 20-20 vision, with or without corrective lenses
- Blood pressure of 140/90
- A height of between 5'4" and 6'3"

Drax's perfect people appeared to be too young to have the necessary education and training to survive life in outer space. Nor did they seem particularly serious people. They were totally preoccupied with glamour and romance—not exactly the space pioneer types.

9

Defeating Supervillains

SPECTRE: Special Executive for Counterintelligence, Terrorism, Revenge, and Extortion. The four great cornerstones of power, headed by the greatest brains in the world.
—Dr. No to 007, *Dr. No*

Every good hero needs a good villain. In the James Bond series, villains are everywhere, but only a few of them are truly memorable. The amoral Auric Goldfinger tops the lists of most Bond fans as the most despicable of all supercriminals. Even Goldfinger's first victim, Jill Masterson, the golden girl, remains etched in our thoughts forever. Standing close behind Goldfinger in the memorable villain category are the DNA-altered Colonel Moon/Gustav Graves; Renard, the man who feels no pain; Max Zorin, the psychopathic genius; and Jaws, the world's most dangerous assassin. How believable, we wonder, are these rogues? Are they scientifically possible or sheer nonsense? And what about painting a woman with gold paint? Will it kill her or just give her a bad case of heat stroke?

Gustav Graves

In *Die Another Day*, James Bond finds himself pitted against the vicious North Korean officer Colonel Moon, who dreams of conquering South Korea and then Japan. When Moon is nearly killed in a fight with Bond, he travels to Dr. Alvarez's mysterious clinic in Cuba, where he undergoes DNA replacement therapy. Exit the Korean colonel Moon, enter the very British-appearing Gustav

Graves, the billionaire owner of a diamond mine in Iceland. James Bond discovers Dr. Alvarez's clinic while tracking down Zao, Colonel Moon's longtime assistant. Bond stops Zao's transformation by gene therapy into a German industrialist, while Giacinta "Jinx" Johnson, an American CIA agent, kills Dr. Alvarez and blows up his hospital.

It's all very twenty-first century, but strictly formulaic. If *Die Another Day* had been made in the 1960s or 1970s, Dr. Alvarez, instead of specializing in DNA replacement therapy, would have been a plastic surgeon. Villains' faces were altered without great difficulty in both *Thunderball* and *Diamonds Are Forever*. DNA replacement therapy sounded more modern, and since it involved using the subject's bone marrow, it was (according to Dr. Alvarez) more painful—enough so that Colonel Moon/Gustav Graves never slept.

Exactly what is DNA replacement therapy, and is it on the brink of replacing plastic surgery? Will we all soon be able to change our features as easily as we change clothes, or are we stuck with what we've got?

First, we need to cover a little background. Genes are the biological building blocks of heredity. They determine our body's traits, including our eye and hair color, as well as the ability of our blood to carry oxygen. More complex traits, such as strength, are shaped by the interaction among different genes as well as by a person's physical environment.

A gene is part of a deoxyribonucleic acid (DNA) molecule. A human being has between 50,000 and 100,000 genes. Genes contain the instructions that allow cells to produce specific proteins. To do this, cells use another molecule, ribonucleic acid (RNA), to translate the information stored in DNA.

Only a subset of genes in a cell are working at any particular instant. As cells grow, various genes become inactive. The pattern of active and inactive genes in a cell and the protein that they produce determine what type of cell it is. Certain cells can do certain things in the human body. Flaws in genes often result in disease.

Scientists are learning how to manipulate and alter genetic

material in people to fight or prevent disease. DNA therapy (or gene therapy, as it is often called) is still in the experimental stage. It involves introducing genetic material (either DNA or RNA) into a person's cells to fight disease. At present, the most common use of gene therapy is in treating cancer.

A gene can't be directly inserted into a patient's cell. It has to be delivered to the cell using some sort of carrier, called a vector. The vectors normally used in gene therapy are viruses, which possess an amazing ability to recognize cells and insert their DNA into them.

In some gene therapy experiments, cells are taken from the patient's blood or bone marrow (as mentioned in *Die Another Day* by Dr. Alvarez) and grown in a laboratory. The cells are exposed to a virus carrying the desired gene. The virus enters the cells and inserts the gene into the DNA. The altered cells are further grown in the laboratory and then injected into the patient. This gene therapy technique is known as ex vivo, as the cells are grown outside the human body. Plus, the gene transfer takes place while the cells are outside the patient's body.

Sometimes, vectors are used to directly deliver the desired gene to cells in the patient's body. This type of gene therapy is known as in vivo, as the gene transfer is done inside the patient's body.

Gene therapy is still a vast new medical frontier and is filled with risks. Using viruses as vectors is a dangerous procedure and must be done carefully. Scientists must deliver genes consistently to precise locations in the patient's DNA. One small mistake and the patient will die.

Today, gene therapy is used to correct various genetic flaws in newborns and to help cure life-threatening diseases. Many doctors and researchers believe that within five to ten years, gene therapy will be a common procedure. Scientists will possess the power to enhance human abilities such as memory and intelligence. Already, some genetic experiments have been done to enhance people's features. Scientists in Germany are working on a new DNA shampoo they feel certain will grow hair. A second version of the shampoo will permanently remove hair from where it's not wanted. The German experiments working with the human genome code are only the tip of an immense iceberg of DNA research being conducted throughout the world.

Dr. Alvarez's clinic in *Die Another Day* doesn't exist—yet. At least we don't think it does. But it probably will in ten years.

Renard

Creating a unique villain isn't easy, as we try without success to remember the names of any of the hordes of forgettable bad guys from hundreds of spy and thriller movies produced during the past few decades. Even the most memorable Bond villains only remain fixed in our thoughts because of their overreaching personalities (Goldfinger) or bizarre physical traits (Jaws). Still, every producer and director tries to come up with the best antagonist he or she can, even if it sometimes means endowing the villain with incredible powers or, in the case of Renard, the global terrorist in *The World Is Not Enough*, with an incredible handicap.

Renard was the name taken by Victor Zokas, a Russian KGB agent, after he was cast adrift by his handler in the Russian spy organization because of his independent nature. After organizing his own terrorist network, he was soon considered by MI6 and other international security agencies to be the world's most dangerous revolutionary. Some years before the events in *The World Is Not Enough*, 009 was sent by M to assassinate Renard. The attempt failed, but only by inches. Renard was left alive, but with a bullet lodged in his brain.

Though the slug in his medulla oblongata didn't kill him, Renard was left unable to feel pain. Little by little, he was losing his other senses. The bullet was killing him slowly, but before dying, Renard dreamed of finishing one final mission: detonating a nuclear warhead in the harbor of Istanbul.

In a rare deviation from form, Renard was actually portrayed as almost human, a failing not normally associated with Bond villains. Usually, the antagonists are multimillionaires seeking to control the world or greedy businessmen wanting to reach the billionaire level. They are evil men, like Ernst Stavro Blofeld or Gustav Graves, without any redeeming factors. Renard, though he fulfilled all the requirements of a bloody terrorist, actually expressed his love for his partner in terror, Elektra King. And he mourned his own inevitable fate instead of displaying the usual stoic demeanor of Bond villains.

Pain is our body's method of warning us of danger. It stops us from repeating any action that causes pain. It warns us that something is wrong. Touching something painful activates an instant withdrawal reflex. Pain tells us when we have been sitting in the same position too long and need to move about to restore blood flow. Pain tells us when we have been injured and need to fix the wound. Pain tells us that we've been working too long and our muscles need a break.

In extremely rare cases, children are born unable to feel pain. Familial dysautonomia is a rare genetic disorder of the autonomic nervous system that primarily affects people of East European Jewish heritage. This condition is very dangerous. Eyes become very dry and eroded because the brain receives no signals to tell the eyelids to blink and send tears to the front of the eye. When the child begins teething, he or she cuts his or her tongue. Joints wear out because there are no pain signals to warn the child to take it easy.

An absurd example of misunderstanding this concept appears in *The World Is Not Enough*. Renard, seeking to prove to one of his subordinates that he is willing to do anything to achieve his goals, grabs a red-hot brick and squeezes it in his hand. He does not, of course, suffer any pain. He then forces his henchman to do the same, which badly mutilates the man's hand. The problem with the scene is that Renard's hand would have been charred and burned just as badly as the other man's; he just wouldn't have felt the pain. Feeling no pain is not the same as experiencing no damage.

People suffering from leprosy lose their ability to feel pain in their arms and legs. Because of this, many of them suffer severe damage to their hands and feet without realizing it. Almost 60 percent of diabetics suffer from peripheral neuropathy, which is the failure of the nerves that carry information to and from the brain and spinal cord. This disorder leads to a loss of sensation in the arms and legs.

Some people can actually teach themselves not to feel pain

through self-hypnosis. Hypnoanalgesia is the inability to feel pain while in a trance by directing attention away from the injured spot. The body, however, is still damaged. No one is exactly sure how this method works.

Thus, Renard's condition, while not hereditary or caused by some disease, was not as rare or unusual as he thought. Had he abandoned his terrorist ways, he most likely could have survived for years with good care and proper treatment.

Explaining Renard's injury isn't extremely difficult, but explaining what the injury does to his brain raises a few problems. According to the movie, Renard's inability to feel pain is the result of a bullet lodged in the medulla oblongata. Recent studies about pain reception in the brain cast doubt on that assertion, however.

Using functional magnetic resonance imaging (FMRI) during the past decade, studies of the brain have identified numerous areas involved in processing pain. Though many of these areas are highly interconnected and do many different things, they normally have been studied on their own. Thus, little attention has been given to the possibility that the brain deals with pain on a much larger scale than believed.

FMRI studies of human brains have identified that a large number of brain areas are activated when pain occurs. These include bilaterally in the cerebellum, putamen, thalamus, insula, anterior cingulate cortex, and secondary somatosensory cortex; contralaterally in the primary somatosensory cortex and supplementary motor area; and ipsilaterally in the frontal operculum. Thus, while the bullet in Renard's brain might have been slowly killing him, it's extremely unlikely it was also blocking him from feeling pain. Most likely, Renard was merely crazy.

Max Zorin

Max Zorin, grandly overplayed as a psychopathic murderer and millionaire by Christopher Walken, was the lead villain in the fourteenth Bond film, *A View to a Kill*. Aided by his deadly hitwoman, May Day, played by Grace Jones in an equally outrageous, over-the-top performance, Zorin seeks to gain a monopoly on the world microchip market by causing twin earthquakes that would sink Silicon Valley beneath the ocean.

Zorin, we learn in bits and pieces throughout the movie, was the result of genetics research conducted by Dr. Carl Mortner in Nazi death camps. Mortner hoped to create a race of genetically superior superchildren by injecting pregnant women with massive amounts of steroids. Unfortunately, all the children born as a result of the experiment became psychopaths. These insane geniuses were dubbed "the steroid kids" by Chuck Lee of U.S. intelligence. At the end of the war, Dr. Mortner and his children disappeared inside the Soviet Union.

Years later, Max Zorin appears in Europe. With the secret backing of the KGB, Zorin builds a huge business empire while passing industrial secrets to his Soviet handlers. It is during this period that Zorin comes up with the idea for Project Main Strike. Dissolving his ties with the Soviets, Zorin schemes to become the richest man in the world by gaining total control over the world's microchip supply. He is stopped by James Bond, however, and permanently put out of commission.

Zorin is a brash, arrogant billionaire with little respect for other people or their lives. When the opportunity arises for him to kill, even his own people, he does so with a smile on his face. He is loyal to no one, condemning to death even May Day, the woman he supposedly loves. Yet strangely enough, Zorin is more real than most people imagine.

In 1935, German scientists synthesized the first anabolic steroid, androstenedione. A year later, the lead geneticist Charles Kochakian determined that this artificial testosterone could be used to increase both muscle mass and mental activity. At the beginning of World War II, Nazi scientists began experiments with anabolic steroids, hoping to increase the aggressive tendencies and stamina of Nazi soldiers.

Soon after, a group of doctors and charlatans began inhumane experiments on prisoners in Nazi concentration camps. Among these madmen was the noted German gynecologist Dr. Karl

Clauberg. Based on reports given by the few survivors of these experiments, Clauberg was one of the few doctors experimenting with artificial insemination. It's quite possible, based on the fact that Nazi scientists were also working with steroids, that Clauberg and his assistant, Johannes Golbel, injected artificially inseminated women with steroids during their pregnancies. After the war, no records of Clauberg's experiments or indicating what happened to the poor women subjected to his mad schemes were ever found. While it seemed unlikely, no one was able to say for sure if these women actually gave birth. Thus, the question was left open that such "steroid kids" actually existed.

At the end of the war, Clauberg was captured by the Soviets. He was tried for war crimes and sentenced to life in prison. He served seven years of his sentence before he died. However, the whereabouts of Golbel were never discovered. It's quite possible that Golbel agreed to work for the Soviets, raising the children born of the cruel experiments. Considering the dangerous effects of steroids on the brain, these children might well have turned into psychopaths. Substitute a few names, change a few dates, and history makes Max Zorin a much more believable villain than he seems.

Jaws

Jaws, as played by the actor Richard Kiel, was a hired assassin standing over seven feet tall who possessed incredible physical strength and titanium-steel teeth. He usually ripped out the throats of his victims with his metal jaws.

Jaws appeared as the secondary villain in two Bond films, *The Spy Who Loved Me* and *Moonraker*. In the second film, he turned on Hugo Drax when Bond convinced the assassin that there was no place for him or his girlfriend in Drax's utopia. Jaws was one of the most menacing and popular villains ever to appear in the Bond saga. In his first appearance, Jaws bordered on the bounds of believability. But in his second outing, the script turned him into a comic-book character who managed to do impossible feats with his teeth.

A normal human being exerts approximately 200 pounds of

pressure when biting on something hard. According to the *1992 Guinness Book of World Records*, in 1986 a man achieved a bite strength of 975 pounds per square inch (psi) for two seconds, the most powerful bite by a human ever. By comparison, an American Bull Terrier, commonly called a pit bull, has a bite strength of 1,200 psi. The alligator, the reptile best known for its jaw strength, has a bite of several thousand psi.

In *Moonraker*, Jaws bites through the cable supporting a cable car traveling above the city of Rio de Janeiro, Brazil. Cable car cables are approximately 11.4 inches in diameter. They are made of six steel strands, each composed of nineteen wires, all of which are wrapped around a sisal rope core. A cable car cable can withstand a grip of 30,000 psi without being damaged. No animal can bite through a cable car wire, not even one with steel teeth. Jaws might have been a fun villain, but his abilities aren't even remotely possible.

Jill Masterson

In an early scene in *Goldfinger*, James Bond, on a visit to Miami Beach, Florida, discovers the evil Auric Goldfinger cheating at cards through an elaborate deception. Goldfinger has his beautiful secretary, Jill Masterson, stationed in a hotel room across from the poolside game. She uses binoculars to read his opponent's hand and relays information to him via a phony hearing aid he's wearing.

Bond puts an end to Goldfinger's winning streak when he finds the attractive Masterson and, in his free time, seduces her. An unexpected visitor incapacitates Bond, however, and when the spy awakens, he finds the beautiful secretary dead, her body completely covered with gold paint. According to Bond, it's death by suffocation, because the human body breathes through pores in the skin. Furthermore, Bond mentions that professional dancers who cover themselves with paint know to leave a small patch of skin untouched at the base of the spine to prevent them from suffering a similar fate.

In the film, it sounded quite logical, and the striking image of Shirley Eaton, the young woman who played Jill Masterson, covered with a veneer of gold paint appeared just about everywhere, from posters to playing cards to the cover of *Life* magazine, promoting the film. The woman turned to gold by the "Midas touch" of Goldfinger was a spectacular ad for the movie. It was one of the most memorable images of the 1960s and seemed to symbolize both the dangerous and the erotic nature of the James Bond series.

According to studio publicity, when the gold scene was filmed, several doctors were present on the film stage to make sure Eaton didn't succumb to the deadly effects of being painted gold. Actually, the actress wasn't in any real danger, since she was wearing a gold bikini bottom, and a six-inch square of her abdomen was left unpainted. The uncolored areas were intended as a safety measure to let Eaton's skin breathe normally. Thus was born an urban legend, that a person could be killed in just a few minutes by having his or her skin covered entirely by paint.

In defense of the innocent moviegoers, it's worth mentioning that in 1964, many people believed that the human body at least partially obtained its supply of oxygen through the skin and that closing off all the pores on a person's body would result in death by asphyxiation. Those same people also held to the belief, as stated by Bond in the film, that by leaving a small part of the body unpainted, nothing would happen. We can only assume that none of the people who believed these notions used a snorkel to swim underwater, where their entire body was completely covered with water for a much longer time than Eaton was covered in gold paint.

Humans are mammals and oxygenate their body by breathing into their lungs. As long as a person breathes through his or her mouth or nose, he or she won't die from asphyxiation. Air enters a person's body through the nose or mouth, passes down the throat to the larynx, and then goes to the trachea, also known as the windpipe. In the chest, the trachea divides into the right and left bronchial tubes, which connect to the lungs. Each lung is approximately 12 inches long. The left lung is divided into two sections, known as the superior and inferior lobes. The right lung is slightly larger than the left and is divided into the superior, middle, and inferior lobes. Separating the two lungs is the mediastinum, which contains the heart, the trachea, the esophagus, and blood vessels.

The bronchial tubes branch into a network of tiny tubes less than .04 inch in diameter. These tubes, called bronchioles, branch into a webwork of even smaller tubes called alveolar ducts. Each of these ducts ends in a group of thin sacs known as alveoli. There are approximately 300 to 400 million of these sacs in each lung. Altogether, they have a total surface area of nearly 1,000 square feet. Each alveolus is surrounded by a huge number of tiny blood vessels called capillaries. Oxygen is diffused from the alveolus to the blood, which then passes through the heart and transfers into the cells of the body.

Since cells constantly need oxygen to work, the lungs are always working. A typical adult takes about fourteen to twenty breaths a minute. The amount of air taken in a single, normal breath is called the tidal volume. For an adult, the tidal volume is about half a quart of air. When filled to capacity, the lungs can usually hold about ten times this amount, or 1.3 gallons. This amount is known as the vital capacity.

Even if Masterson thought she was going to die because she was painted gold, it didn't matter. A person can stop breathing for a short time or alter the rate of breathing, but it's impossible for anyone to voluntarily stop breathing forever. Breathing is an involuntary action, similar to a heartbeat, that is controlled by the nerve centers of the brain stem.

The only problem Masterson might have suffered from being covered with gold paint was that the liquid would have clogged her pores, which would have prevented her from sweating. If she had engaged in some serious exercise or aerobics after her paint job, she may have overheated, although most likely she would not have died. For the record, numerous people in modern times cover themselves with body paint and suffer no ill effects, other than reporting it's a cold way to travel unless you're on the beach. Most body painters are well aware of the Jill Masterson urban legend, which they have unsurprisingly dubbed "the Goldfinger myth."

Despite great effort, we've been unable to find one incident that could have convinced Ian Fleming that the paint story in *Goldfinger* was true or had any basis in fact. One Internet fan site mentions that Fleming was inspired by the death of a Swiss model who was accidentally covered by paint, but we've been unable to track down the story. After much searching, we've reached the conclusion that it's vaguely possible the infamous Buddy Ebsen Tin Man episode from *The Wizard of Oz* may have served as Fleming's inspiration (along with the legend of King Midas and his golden touch).

In October 1938, principal photography began on the movie version of *The Wizard of Oz*. Originally, Buddy Ebsen was cast in the role of the Tin Man. Costuming Ebsen to look like a mechanical man proved difficult, and different methods were tried, including using silver paper, silver cloth, and real tin. Finally, the costume department settled on using aluminum dust sprayed over white clown makeup. Everything seemed fine until Ebsen began to suffer from shortness of breath, and his fingers and toes started cramping.

Late one night, after approximately a week on the set, Ebsen woke up at home in the midst of terrible muscle cramps that traveled up and across his body and caused his lungs to collapse. The aluminum dust in the makeup caused an allergic reaction that put Ebsen in the hospital for two weeks and out of *The Wizard of Oz*.

Ebsen almost died because of his reaction to his makeup. It's possible that Fleming read the story and came up with a golden twist for his novel. Whatever the inspiration, the painted golden girl of *Goldfinger* has become one of film's greatest urban legends. As we mention elsewhere in this book, however, it's not the only myth that the movie created.

10

Nullifying the Threat of Superweapons

*If we blow up Kansas, the world may not
hear about it for years.*
—ERNST STAVRO BLOFELD, *DIAMONDS ARE FOREVER*

One of the most common plots in 1930s spy novels was the development of a new superweapon that would change the face of warfare forever and the desperate struggle by the hero to destroy that weapon before it could be used to eradicate the fragile peace of the time. The weapon of choice was usually some sort of death ray that could kill soldiers from a long distance without the inconvenience of destroying surrounding property. Thus, in book after book, mad scientists working for unnamed governments (though all the plotters had names like Hans and Fritz and spoke with thick German accents) schemed to loot the Louvre and dismantle the Eiffel Tower in an otherwise deserted Paris. Until, of course, their dreams ended in spectacular explosions when Simon Templar (the Saint), Bulldog Drummond, or one of a half-dozen other adventurers of that era put an end to the mad schemes. The explosions also killed the evil masterminds, their diabolical henchmen, and the only existing plans for the superweapons, thus guaranteeing safety and freedom for all of Europe until the rise of a new villain a few months later.

For the most part, the James Bond adventures avoided the superweapon cliché. Auric Goldfinger and Emilio Largo relied on nuclear weapons to promote their evil plans. Carl Stromberg, Ernst Stavro Blofeld, and Elliot Carver tried to manipulate the major nuclear powers into atomic war through intrigue and deception. In all cases, while the villains worked with technologically advanced machinery, they still required atomic bombs for the success of their schemes. Others planned attacks using biological weapons: deadly but usually impractical, since they were not easy to control and their effects were unpredictable.

Only a few madmen were willing to use new weapons of mass destruction. It took men of courage and near insanity to attack the world with an electromagnetic pulse cannon, as did Alec Trevelyan in *GoldenEye*; to try to sink part of California into the ocean, as Max Zorin did in *A View to a Kill*; or to melt away obstacles in your path with a giant space laser, as Gustav Graves did in *Die Another Day*. Villains such as these men give us hope for the future of the James Bond film series.

Still, we're forced to ask the usual all-important questions: Are such weapons scientifically possible, even probable in the next few years? Or are they entirely the products of some screenwriter's keyboard?

The Most Dangerous Weapon Never Seen

GoldenEye was the seventeenth James Bond movie. It appeared in 1995, six years after *Licence to Kill*, the second and final film to feature Timothy Dalton as Bond. The long break between films was the result of a complicated legal battle with the series' producers and United Artists over the control of the movies. In the meantime, Dalton left the series and Pierce Brosnan took over the lead role. Anxious to thrust the Bond franchise back in the public eye, the producers pulled out all stops on the film, filling it with incredible action sequences and a complex plot that defied the boundaries of logic or sense. The lack of a coherent story line didn't stop the movie from earning $106 million in the United States and over $350 million worldwide.

The adventure begins with a flashback to a mission years earlier.

James Bond and Alec Trevelyan penetrate a secret Russian chemical weapons plant being run by the diabolical Russian general Arkady Grigorovich Ourumov. The two British agents manage to sabotage the facility, but Trevelyan is captured, while Bond escapes. We zoom to the present, where after some minor shenanigans, Bond witnesses a beautiful but deadly spy, Xenia Onatopp, stealing a new Tiger helicopter, a flying machine invulnerable to electronic attack.

The story switches to the Russian Space Weapons Control Center in Severnaya, Russia. (The real Severnaya is in northern Russia, but a map in the movie places it in the middle of the country for some unknown reason.) With the help of the computer hacker Boris Grishenko, who works at the base, General Ourumov and Onatopp kill everyone in the center. Only Natalya Fyodorovna Simonova, a beautiful computer programmer, survives. Ourumov steals the computer key for the Russian space weapon GoldenEye (which was also the name of Ian Fleming's estate in Jamaica) and he, Onatopp, and Grishenko depart right before the weapon is activated.

GoldenEye is a space-based electromagnetic pulse (EMP) weapon. It fires an EMP wave at the base and fries all electrical circuits within a 30-mile radius of the target. Evidently, the space center is so reliant on electricity that steel girders collapse, fires break out, and all sorts of major structural damage occurs. Still, plucky Simonova manages to escape the destruction and heads for Moscow.

Bond is assigned to destroy GoldenEye. He immediately travels to the Kremlin, where he encounters Simonova under some less than pleasant circumstances. Still, Bond, being Bond, manages to escape his captors, save the girl, and kill the evil General Ourumov. What he doesn't succeed in doing is capturing his old buddy Alec Trevelyan, who is behind the entire GoldenEye operation, along with his assistants, Onatopp and Grishenko.

Bond manages to track the evil trio to Cuba, where he learns that a second GoldenEye satellite orbits Earth. Trevelyan's master plan calls for Grishenko to hack into the Bank of England and electronically steal billions of dollars. Immediately afterward, the second GoldenEye will fire an EMP blast at London, wiping out all electricity in the city and plunging it into chaos. By destroying all electronic records from the past decade, Trevelyan will be assured that Grishenko's robbery will remain undiscovered and that he and his team will become very rich.

By now, Bond fans realized that *GoldenEye* owed much of its plot to *Thunderball*, but the gaps in logic were more severe than before. Why there were two GoldenEye satellites was never explained. Nor did it seem very believable that British and U.S. intelligence never knew about the EMP weapons. Not to mention that destroying all of London to cover up a robbery, even one amounting to billions, seemed rather overindulgent, though Trevelyan did have other issues with British intelligence.

As in all Bond movies, the villain is much too smug. He has the opportunity to kill 007 immediately on capturing Bond when he enters the secret base, but doesn't. The chance passes, explosions occur in the underground computer center, the GoldenEye satellite plunges into the atmosphere and burns up, and Trevelyan and his followers are all killed. Bond gets the girl and the movie ends, leaving a lot of unanswered questions, mostly about EMP weapons.

GoldenEye was the first of a number of films to employ EMP weapons. The electrical nullifier device appeared in *Escape from L.A.*, *Ocean's Eleven*, *The Core*, *Eraser*, and most notably in the three *Matrix* movies, where it was the last line of defense for beleaguered humanity fighting intelligent robots. The only problem with all the films was that they glossed over the principles of the weapons and never really explained how they worked, other than with vague remarks such as "it knocks out all electric circuitry."

The theory of electromagnetic bombs (E-bombs) comes from the work of the physicist Arthur Holly Compton in 1925. Compton showed that shooting a stream of energized photons into atoms with low atomic numbers caused those atoms to eject a stream of electrons. This occurrence became known as the Compton effect.

Scientists never associated the Compton effect with atomic bombs until one fateful day in July 1962. On that day, Test Shot Starfish took place over Hawaii. A hydrogen bomb was detonated about 650 miles out in space. The blast released huge bursts of gamma rays (which served as photons), which upon striking hydrogen and nitrogen in the atmosphere released a vast wave of electrons (as predicted by the Compton effect) that spread into the atmosphere. This gigantic pulse of electromagnetic energy was similar to radio microwaves. The energy of this pulse was so powerful, however, that it erased magnetic memories and melted the microscopic junctions in transistors over 2,000 miles away from the blast. This tidal wave of electromagnetic energy was invisible to humans but powerful enough to blow out light bulbs in the northeast while tripping garage-door openers in Hawaii. Thus began the ongoing saga of the unstoppable EMP weapon.

After this first and only demonstration of the power of electromagnetic waves, the U.S. military set out to learn how to protect our weapons systems from a possible E-bomb. Equal emphasis was placed on developing an E-bomb that didn't require a nuclear device to set off such an explosion. The hunt began in 1962 and continues today. According to various reports issued by numerous U.S. government agencies in the 1960s, 1970s, and 1980s, E-bombs were always on the verge of being invented. Research in other nations such as Russia, Libya, Germany, France, Iraq, and Iran was intense, and the need existed for the United States to keep pace in the EMP race. It all sounded very familiar, because it was. Along with the nuclear arms race, our nation engaged in a secret and rarely discussed EMP race.

In a 1997 paper for *Military Intelligence* magazine, Major Scott W. Merkle used declassified government documents to weave together a strong argument that EMP weapons really exist but are hidden from the public by a worried government. According to Merkle, a declassified U.S. military study[1] claimed that an explosion of a 1-megaton bomb 800 miles over Omaha, Nebraska, would produce an EMP wave that would disable every computer in the United States, southern Canada, and northern Mexico. The same report warned of a possible nuclear attack by Iraq in revenge for Operation Desert Storm that would cause an EMP disaster in the United States. Fortunately, as we discovered during the invasion of Iraq,

Saddam Hussein had no megaton nuclear bombs or any method of launching one to a position 800 miles over Omaha.

Merkle also discussed a report issued by the French National Assembly in 1992 that recommended that EMP weapons should be studied during France's underground nuclear tests in 1993.[2] Merkle further claimed that "weapons designers specializing in high-energy physics can now create electromagnetic pulses without going into outer space. One approach involved harnessing the force of a conventional explosion. Others were simply just modifications of radar, which bounced pulses of energy off aircraft in flight."[3]

Unfortunately, while the Merkle report was one of the most detailed studies of EMP weapons up to that time, it offered no actual proof of the existence of the weapons. Merkle's study did mention how in 1993 several cars parked 1,000 feet away from an experimental EMP generator at Eglin Air Force Base in Florida were discovered to have malfunctioning alternators. He blamed the problem on an unauthorized test of the EMP generator. His evidence was not overwhelming.

According to an article in *Law Enforcement News* on September 30, 1996, the U.S. Army and the National Institute of Justice spent $500,000 on field tests to see if EMP guns could force a car to stop by disabling its electrical system. The police plan was to put an end to high-speed car chases by hitting the fleeing vehicle with an EMP blast.[4] How this blast was going to work without disabling every other car on the highway was never discussed, and no such plan ever was implemented. The strange, nonverifiable world of EMP weapons was getting stranger and more unverifiable with each report.

The greatest problems facing EMP weapons are size and power requirements. To create an actual EMP, the force of an exploding atomic bomb is needed. Despite Internet rumors to the contrary, small atomic devices are not readily obtainable by anyone with a satchel full of money. Even with a bomb, focusing the EMP wave is not a sure thing. Aiming gamma waves is not a job for any ordinary sort of cannon. To get around these problems, modern-day inventors take the easy way out: they ignore them.

A second generation of EMP weapons was introduced to Congress in 1998 by David Schriner, a former civilian electrical engineer and sometime inventor. Schriner testified before a congressional economic committee that he had assembled approximately $400 worth of old automobile parts in his basement into a much smaller but highly dangerous EMP weapon. His device did not require a nuclear explosion for power. According to Schriner, his EMP weapon could make a car's engine rumble 50 feet away from his lab.

Congress was so impressed by Schriner's EMP device that it awarded Schriner a million-dollar grant to see what he could build with "real" money. Using the funds, Schriner set up a laboratory, where he conducted experiments using ordinary hardware-store items to create weapons of electrical destruction. According to one of Schriner's reports, he used a telephone transformer block to construct an EMP device that could scramble medical equipment in a hospital. Another machine, listed as in the development stages, was an EMP broadcast tower that if placed in a van and driven on Wall Street, could shut down the New York Stock Exchange.[5]

Taking Schriner's claims as fact, in February 1999 Diane Sawyer, on the ABC news program *20/20*, interviewed the inventor and witnessed a demonstration of his EMP weapon. After the two donned protective suits and copper mesh helmets, Sawyer watched Schriner use his original device on her Corvette and a donated limousine. Aimed at the open engines from approximately 10 feet away, the EMP gun made the idling engines run rough. Sawyer later claimed the car's electronic door locks also moved up and down. No mention was made of the fact that the electronic video cameras located inside the cars continued to function perfectly during the entire attack.[6]

Reviews written in the months following the show claimed that Schriner's device completely destroyed the motors in the two cars, savaged a number of computers, and nearly destroyed the set. As the story got told and retold, the details grew increasingly lurid. One fact that always remained the same was Schriner's claim that he built his EMP weapon using $400 worth of parts from Radio Shack.

During the same *20/20* broadcast, Sawyer stated that criminals in Russia had used EMP guns to disarm a bank security system. The only problem with the story was that it was based on phony news posted on *Crypt Newsletter*'s Web site in 1997.[7] In the months that followed, *20/20* never acknowledged the mistake.

Not to be outdone by *20/20*, on October 19, 2001, ABCnews.com ran an article titled "E-Bombs Could Spell Digital Doomsday," by Paul Eng. The piece reported the imminent dangers of EMP weapons, such as the hottest new product in electronic warfare: the flux compression generator (FCG). According to a nameless spokesperson from the U.S. Army, an FCG could be assembled from commonly available electronic parts for less than $1,000.[8]

Within a short time after the ABC report, diagrams for constructing an FCG, labeled by Internet gurus as a small but very dangerous model of an EMP weapon, appeared across the Internet. These sites warned that building the weapon was probably illegal and possibly dangerous, but then proceeded in detail to describe how to assemble it. Again, the cost of building an FCG was listed as less than $400, with all of the necessary parts available at Radio Shack.

According to the *Popular Mechanics* Web site, in the article "E-Bomb" by Jim Wilson, an FCG would produce "a peak current of tens of millions of amps, a pulse that . . . makes a lightning bolt seem like a flashbulb in comparison."[9] The results listed were always theoretical, since no one seemed to have actually tested an FCG. In the same article, Wilson quoted an unnamed U.S. Air Force spokesman who likened the FCG pulse to a lightning bolt and stated that electronic devices could be protected by using Faraday cages, which divert EMPs to the ground. According to Wilson, equally unnamed "foreign military analysts say this reassuring explanation is incomplete."[10]

The most amazing result of Sawyer's story on *20/20* is that it created an urban legend that has grown more incredible with each retelling. Taken as truth, it's a story that makes the plot of *GoldenEye* seem tame. In reality, it serves as a powerful reminder of what can happen when journalists run amok.

Several weeks after Schriner's February appearance on *20/20*, *Newsweek* magazine, in its Monday, March 15, 1999, issue, ran a story on the supposed theft of nuclear weapons technology from Los Alamos Laboratory, New Mexico, by Chinese agents. Along with stealing data about the Trident II missile's W-88 warhead, the Chinese were accused of taking information about possible EMP weapon tests. As this was the first and only time EMP tests were mentioned in connection with Los Alamos, it seemed likely that the mention of the weapons relied more on the *20/20* report than on any government statement. This was especially possible since, as determined by a later investigation, it turned out that the entire technology theft story wasn't true.

On Friday, March 19, 1999, four days after the *Newsweek* story appeared, President Bill Clinton held a news conference at the White House during which he fielded a variety of questions from reporters. While Clinton was prepared for the inevitable Monica Lewinsky questions, he was caught by surprise when Wendell Goler of FOX News asked him about the EMP story. According to Goler, unidentified sources told FOX News that Chinese agents stole the technology for EMP weapons from four nuclear labs during Clinton's first term in office. Goler also claimed that the same sources said that the Chinese had successfully tested EMP weapons in China.[11]

When Goler asked Clinton what he was going to do about the report, Clinton made it clear that he doubted the entire story and questioned the reliability of the undisclosed sources. No matter. The next day the *New York Times* reported that an unnamed U.S. official said that intelligence reports indicated that China was certain that it had obtained the technology necessary to develop an EMP gun.[12]

In the weeks that followed, several hundred Web sites picked up this story as an indication of how President Clinton had compromised U.S. security by letting China steal the plans for EMP weapons. Once reported in the *New York Times*, this science story attained a certain recognition that no denial could ever erase, not that the *Times* ever published a correction concerning China's EMP guns, although other periodicals did.

On September 17, 2002, Major General Tyson Fu of the National Defense University of Taiwan called newspapers unprofessional for raising concerns that China might use an EMP weapon against Taiwan, since such a weapon had never been shown to exist. According to published reports, the general said that it was incredible that the press had made an issue out of a nonexistent weapon.[13] The general specifically cited *Newsweek* as the main source for EMP rumors; maybe he didn't subscribe to the *New York Times*.

In an interview with *Crypt Newsletter*, Neal Singer of Sandia National Laboratory, one of the government centers researching EMP weapons, called EMP weapons "an interesting urban legend." According to Singer, the military wasn't interested in EMP devices because the power needed to generate EMP effects would be so destructive that anyone using the weapon would be killed or seriously injured.[14]

In the ten years since *GoldenEye* first appeared on the screen, reports of EMP weapons and FCG devices have increased a thousandfold. There have been numerous congressional hearings on the possibility of an electronics-only war. Dozens of reports claiming that the U.S. Air Force used EMP bombs in the invasion of Iraq have been published in everything from newspapers to news magazines to tabloids. When Iraqi TV went off the air during the invasion, a number of security specialists immediately suggested that an EMP bomb had been used. Hours later, the air force revealed that an ordinary bomb had been responsible for the damage.

Despite not one verified EMP attack, rumors continued to circulate about the unseen menace. Magazine articles quoted unnamed sources in the Pentagon leaking stories of EMP bomb tests in the early 1990s. A British firm announced that it had developed an EMP bomb operating on radio frequencies, but for some unexplained reason the British government didn't seem interested. Alleged conversations between high-ranking Russian generals and U.S. senators were reported, describing the havoc that would be created by their E-bomb arsenal, but the names of those talking were never revealed. New congressional committees called the same witnesses, engineers, and scientists who had been interviewed by previous committees to testify, and their testimony was treated as new. Old reports, based on rumors and even older reports, were entered into the record as fact. Misinformation was piled on misinformation and reported to the public as fact.

After searching for the truth, we finally conclude that while a lot has been said about EMP weapons, especially to congressional committees; much has been written about EMP weapons by correspondents from *Time*, *U.S. News & World Report*, and the *New York Times*; and there are hundreds of rumors about secret demonstrations and covert attacks using EMP weapons all over the Internet, the damned things do not exist. Moreover, EMP weapons do not exist at present, did not exist when they were first proposed in the past, and most likely will not exist for years to come. The only reason there are so many stories about them is that no one is interested in the truth when the urban legend is so fascinating.

Sinking Silicon Valley

There's little disagreement among Bond fans that *A View to a Kill* was one of the worst Bond disasters ever filmed. Even the incredible overacting by Christopher Walken and Grace Jones wasn't enough to save this disjointed, totally outrageous film that was Roger Moore's swan song to the series. At fifty-eight, Moore looked terribly out of place and quite exhausted, fighting the genetically enhanced Max Zorin on the top of the Golden Gate Bridge. It was clearly time for a younger actor to carry on the Bond legacy.

The film's plot, ripped right from the pages of *Goldfinger*, did nothing to ease Moore's pain. Zorin, a genetically enhanced supergenius funded by the Russians, controls much of the world's supply of microchips. All of his European and Asian rivals know they must sell their chips through his corporation. America's Silicon Valley dominates the microchip industry, however, and Zorin can't compete with the best-made chips in the world. So he decides to destroy them.

The basic premise of *A View to a Kill* is that Zorin wants to become the richest man in the world. He plans to do this by setting off explosions on the San Andreas Fault, causing an immense earthquake and sinking Silicon Valley into the Pacific Ocean, thus leaving him the world's only major supplier for computer microchips.

First, we must point out that the world's microchip market is controlled more or less by Intel, and it managed to do this without causing any earthquakes. Sometimes, things accomplished by violence can be done just as easily without a shot being fired. The easiest way for a company to corner the market is to produce the best product and make sure it is available at the cheapest possible price. No other solution works half as well.

Second, we need to mention that in 1999, a major earthquake hit Taiwan, causing a serious disruption in semiconductor production in Asia. Businesses did not collapse or implode, however. Instead, they found alternative sources of supply. Life in the electronics industry goes on. The market has become too globalized for any one disaster to slow down technology.[15]

Still, we can't help but wonder if Zorin's plan might have worked if Bond, with the unexpected help of May Day, hadn't saved the West Coast at the last possible instant. The unsettling answer is maybe. In 1968, U.S. earthquake experts discovered that controlled nuclear underground explosions in the Nevada desert were exerting seismic stress on the earth's surface. One blast is thought to have created a new fault line in the earth's crust nearly three-quarters of a mile long.[16]

Studies of seismic waves generated by a nuclear blast show that up to one-third of the explosion's force exerts pressure with earthquakelike seismic movements. A 5-megaton atomic explosion would release seismic waves that measure 6.9 on the Richter scale. Unlike normal earthquakes, however, nuclear detonations don't produce waves that travel very far from the center of the blast. At best, they travel a thousand feet underground. Still, a few well-placed nuclear bombs in the San Andreas Fault could mean a major disaster for Silicon Valley.[17]

Needless to say, the movie received fair reviews in 1985 and was soon forgotten—until 1989.

The Loma Prieta earthquake occurred on October 17, 1989, in

the greater San Francisco Bay Area in California at 5:04 P.M. It measured 7.1 on the Richter scale and lasted for fifteen seconds. Its epicenter was about 10 miles northeast of the town of Santa Cruz, near Loma Prieta Peak in the Santa Cruz Mountains. The earthquake caused severe damage as far as 70 miles away, including San Francisco, Oakland, and the San Francisco peninsula. Severe damage occurred closer to the epicenter, in the communities of Santa Cruz, Watsonville, and Los Gatos.

The magnitude and distance of the earthquake from the severe damage to the north were surprising to geologists. Studies done since the earthquake indicate that much of the damage was due to reflected seismic waves.

There were 63 reported deaths and 3,757 reported injuries as a result of this earthquake. The highest concentration of fatalities, 42, occurred in the collapse of the Cypress structure on the Nimitz Highway (Interstate 880), where a double-decker portion of the freeway collapsed, crushing the cars on the lower deck. One 50-foot section of the bay bridge also collapsed, causing one car to fall to the deck below and resulting in the only fatality on the bridge. The bridge was closed for repairs for about a month and reopened on November 18.

The quake also caused an estimated $6 billion in property damage, the most costly natural disaster in U.S. history at the time. It was the largest earthquake to occur on the San Andreas Fault since the great San Francisco earthquake of April 1906.[18]

Silicon Valley is the nickname for the southern part of the San Francisco Bay Area. It includes the northern part of Santa Clara Valley and adjacent communities in the southern parts of the San Francisco peninsula and East Bay. It stretches approximately from Menlo Park and the Fremont/Newark area in the East Bay down through San Jose, centered roughly on Sunnyvale.

While most of Silicon Valley was spared from the Loma Prieta earthquake, the disaster made it clear that the area was extremely vulnerable to a similar event. Still, it wasn't until ten years later that a meeting was held at Compaq Computer Corporation's West Coast headquarters to discuss earthquake problems in Silicon

Valley. This 1999 meeting brought earthquake experts, risk managers, and geographic information system specialists from private industry together with local, state, and federal government agencies to develop a comprehensive HAZUS earthquake risk assessment for the San Francisco Bay Area. HAZUS was developed by the Federal Emergency Management Agency (FEMA) as a nationally applicable software program to estimate potential losses from earthquakes, floods, and hurricanes.[19]

The conference was the first national meeting to acknowledge that America's computer industry might be resting on shaky ground, just as Ian Fleming predicted nineteen years earlier in *A View to a Kill*.

Lasers in Space

As with many of the Bond films, the plot of *Die Another Day* isn't strong on logic. At the opening of the film, our hero sneaks into North Korea and takes the place of an Australian conflict diamond merchant. Conflict diamonds come from regions controlled by revolutionary forces opposed to legitimate governments, usually in Africa, and are used to fund military action against those governments. The sale of conflict diamonds is illegal under international law.

The merchant is about to sell the diamonds to the evil Colonel Moon in exchange for weapons. The weapons that Moon and his henchman, Zao, are exchanging for the jewels are huge hovercraft that can skim over the many thousands of land mines buried in the demilitarized zone (DMZ) between North and South Korea. Exactly why Moon is trading useful weapons for diamonds isn't explained, nor does it make any sense that a diamond dealer (or arms dealer) would want to buy the hovercraft, which are useful only in places like North Korea.

Unfortunately for Bond, during the deal his true identity is revealed. He seemingly kills Colonel Moon by pushing him and his hovercraft off a cliff but then is captured by the North Koreans. Fourteen months of torture pass before he is returned to the British in a trade for Zao. M thinks Bond cracked under intense pressure, so she refuses his request to be allowed to track down the person who betrayed him. Bond does so anyway, revealing a traitor in the British Secret Service. More important, he discovers that Colonel Moon is alive, disguised as the billionaire Gustav Graves. Moon is still dealing in illegal diamonds, which he is passing off as legitimate stones found in a diamond mine in Iceland. It's never made clear exactly what he gives to the Africans to get the conflict diamonds, since he keeps half the money from their sale and donates the rest to charity. The funds raised by selling the diamonds have been used to construct a gigantic solar laser called Icarus. Moon pretends he is going to use Icarus for peaceful purposes, but he actually intends to use the superlaser to destroy all the land mines in the DMZ between the two Koreas. Once the minefield is cleared, Moon intends to lead the North Korean army in an invasion of South Korea and, after that, Japan.

In one of the most exciting climaxes of any James Bond film, *Die Another Day* features Bond and his American counterpart, the female superspy Giacinta "Jinx" Johnson, battling the diabolical Colonel Moon and his various henchmen (and women) on a giant airplane flying over the DMZ. The twin fights on the general's airplane are exciting, the dialogue sparkles with the usual Bond wit, and the imminent danger of the giant laser ray is breathtaking. The only problem, unfortunately, is that anyone watching the film sooner or later asks why Colonel Moon is using the huge space laser to detonate the land mines. Wouldn't it be much easier to focus the laser on Seoul and burn the South Korean capital to the ground? Why bother sending hundreds of thousands of men into battle when you already control the ultimate doomsday weapon?

In the end, Bond wins, Colonel Moon is shredded into confetti, the laser is shut off just in time, and our brave hero and Jinx debate turning in a stray box of conflict diamonds. It's a great show, as long as you don't ask questions, because once you start asking, the movie stops making sense.

After devoting much of this chapter to *GoldenEye* and an EMP weapon that doesn't exist, can we ignore Icarus, a laser in space that can detonate land mines and set a gigantic plane on fire? Of course not! EMP guns, as we discovered, have never actually been demonstrated in public. Laser guns, however, are real. Lasers are used in everything from DVD players to eye surgery. They're part of modern civilization. Why can't we build one the size of Icarus and wipe out terrorist camps across the world? If only life were so easy.

To fully understand the problem, we need to investigate on a fairly simple basis how lasers work and what they can and cannot do. It's actually quite easy if we start at the beginning, with the building block of the universe: the atom. (To be honest, we should mention that atoms are no longer considered the smallest particles of matter in the universe. Quarks, for example, are much smaller, as we discuss in our book *The Science of Supervillains*.)

Atoms make up all matter. They consist of three particles: protons, neutrons, and electrons. The protons and neutrons form a nucleus at the center of an atom. Depending on the size of the atom, a number of electrons spin around the nucleus in orbits. These orbits don't always follow an exact path, but each electron is always a specific distance from the nucleus.

An atom is normally at its ground-state energy level. When we apply energy to an atom through electricity, heat, or light, the atom changes into an excited state. The energy added to the atom is transferred to the electrons, which move from their original low-energy orbit to a higher-energy orbit, farther away from the nucleus. After remaining at a high-energy orbit for a short time, however, an electron usually drops back to its ground-state energy level. When that happens, the electron releases the energy it has absorbed as a photon—a particle of light. Whenever we turn on a lightbulb and induce electrons to rise to a higher-energy level, then drop back to the ground level, photons are released in the form of light.

A laser is an object that controls the way electrons of energized atoms release their photons. *Laser* stands for "light amplification by stimulated emission of radiation." It's quite easy to understand how a laser works once we know the four basic units that make up almost all lasers.

The *lasing medium* is a substance that emits light in all directions. The lasing medium can be a gas, a liquid, a solid, or a semiconducting material. The *excitation mechanism* of a laser is a source of energy that's used to excite the lasing medium. Most excitation mechanisms use electricity from some sort of power supply such as a light or a flashlight. The third part of a laser is the *feedback mechanism*, which reflects the light from the lasing medium back into itself. Usually, this mechanism consists of two mirrors, one at each end of the lasing medium. As the light is bounced between the mirrors, it increases in intensity. The fourth part of a laser is the *output coupler*. This piece is a partially transparent mirror at one end of the lasing material (and forms part of the feedback mechanism), which allows some of the light in the medium to leave. The light that leaves the lasing material forms the laser beam.

Using these four terms, here's the simple explanation of how a laser works: A lasing medium is "pumped" with energy from the exciting mechanism. This extra energy is absorbed by the atoms in the lasing medium, and those atoms reach an excited state. In other words, these electrons move from a low-energy orbit around the nucleus to a higher-energy orbit in the atom.

Shortly after the electrons reach this higher-energy orbit, however, they drop back to a lower-energy level. As this happens, the electrons release photons, a form of light energy. These photons have a very specific wavelength. Two identical atoms with electrons in the same state release photons with the same wavelengths. These wavelengths have a specific color. Thus, all these photons are the same color, producing monochromatic (one-color) light (unlike sunlight, which is composed of many colors).

Some of the photons released in the lasing material run parallel to the center axis of the material, so they hit the feedback mechanism mirrors that are placed at the end of the material. The photons bounce back and forth between the two mirrors. As they bounce, they stimulate the other atoms in the lasing material to release photons of the same wavelength. The number of photons bouncing back and forth against the mirrors increases. All the photons moving together form a monochromatic, coherent, and very directional beam of light. Part of that light leaves the lasing material through the outlet coupler. The exiting light beam is a laser.

Laser light is different from normal light in three primary ways. Ordinary light travels in numerous directions from its source. The light from a laser bounces between the mirrored ends of the lasing medium many times before being released, narrowly focusing the electrons so that they become nearly perpendicular to the mirrors. Thus, laser light emerges in an extremely thin beam in a specific direction. This laser light is said to be highly collimated (having parallel rays).

Second, the light emitted from a laser is monochromatic, as discussed earlier. It's of one color, or wavelength. Third, the light from a laser is coherent, meaning that it is a beam of light whose photons all have the same optical properties: wavelength, phase, and direction. Ordinary light can be a mixture of many wavelengths.

Because of these three properties, laser light can transmit a vast amount of energy in a small area. The highly collimated beam of a laser can be focused to a microscopic dot of extremely high-energy density. Such beams can cut through steel.

Lasers are dependent on the lasing material used to generate the light photons that make up the light beam. A solid-state laser uses a solid such as a ruby as the lasing material. Gas lasers use a gas for the lasing medium. The most common laser is a helium-neon gas laser, which emits red light. Carbon dioxide lasers emit energy in the infrared spectrum and are invisible to the human eye. Carbon dioxide lasers are among the most powerful lasers and are usually used for cutting beams. An excimer laser (a combination of the words *excited* and *dimers*, which are molecules consisting of two identical, simpler molecules) uses reactive gases such as chlorine mixed with an inert gas such as krypton. These lasers produce light in the invisible ultraviolet range.

Dye lasers use complex organic dyes in liquid solutions as the lasing solution. Semiconductor lasers are very small and use low power. They sometimes are known as diode lasers.

Lasers are also characterized by the duration of emitted laser light. A continuous wave laser emits a steady beam of light. A pulsed laser emits laser light in an on-and-off, or pulsed, manner. A Q-switched laser is a pulsed laser with a shutter on the output coupler. Energy is built up in a Q-switched laser and released by the shutter to produce a single intense laser pulse.

Armed with our knowledge of the various parts of a laser and how lasers work, we can now take a look at Colonel Moon's giant solar laser and see if it makes sense. But before we do that, just to be thorough, let's check out James Bond's first close encounter with a laser, courtesy of Auric Goldfinger.

On May 16, 1960, Theodore Maiman at Hughes Research Laboratories invented the first laser, using a ruby as his lasing medium. Maiman's work was based on the studies of Charles Townes at the University of Maryland and Alexander Prokhorov at Lebedev Laboratories in Moscow, both of whom were awarded the Nobel Prize in 1964. While the first laser was a tremendous technological breakthrough, it actually wasn't very useful. The laser beam wasn't powerful enough to be a weapon, as envisioned by the military. Nor could it transmit information through the atmosphere because of its inability to penetrate clouds and rain. It wasn't long before scientists developed lasers that could be used in surgery, however, where a powerful point of heat was necessary.

Still, in 1964, when *Goldfinger* was released, lasers were exotic devices that had never appeared in film. For most moviegoers, the laser weapon that threatened James Bond was the first time they had ever seen or heard of a laser. The bright red ray cutting the steel platform to which 007 was bound was a new and extremely menacing weapon. Only a few people realized that the color of the laser beam indicated that Bond was being threatened by a helium-neon laser, a beam about as dangerous as a ray of sunshine. The steel wasn't actually being cut by the laser at all, but by a technician beneath the table wielding a blowtorch.

If Goldfinger really wanted to bisect Bond, he would have used a carbon dioxide laser, which is quite capable of cutting steel plates with pinpoint accuracy. Unfortunately for filming, a carbon dioxide laser is invisible to the human eye, making for less-than-spectacular viewing. It's well worth noting that carbon dioxide lasers were not used in industry to cut steel until 1966, two years after *Goldfinger* was made. Sometimes the best predictions are the ones not intended to be made.

Moving forward thirty-eight years brings us to Colonel Moon and his space-based laser aimed at the DMZ. (Sorry, but the laser weapon in *Diamonds Are Forever* is so outrageous that we refuse to consider that it is a believable threat.) According to the movie, Graves has placed a giant satellite in space called Icarus, which in reality is some sort of space laser. It can't be a solar laser, because solar lasers use sunlight gathered by solar sails as their excitation mechanism and are thus dependent on the amount of sunlight available while the satellite is traveling around the globe. While the amount of heat might be enough to excite the atoms in the lasing medium, the laser ray generated by the solar laser is much more dependent on the lasing medium than on the energy source. Solar lasers therefore aren't considered dangerous to any objects on Earth.

To create a truly dangerous space satellite at this point in history, Graves would have needed a carbon dioxide laser in orbit, which would have required an orbiting space station and taken numerous technicians several years to assemble. It's never explained exactly how Graves gets his laser into space, but it's obviously been placed there by the space shuttle or a Russian spaceship, which would have been highly unlikely. Nor is it explained what lasing medium is used to produce such a powerful ray.

A satellite weapon capable of producing a laser beam strong enough to punch a hole through a hundred miles of atmosphere, blocked by clouds, rain, and wind movement, would require something much larger than the current airborne chemical laser being developed by the Pentagon for a Boeing 747 airplane. The plane needs to carry huge tanks of oxygen and iodine, as well as a 14,000-pound nose turret to aim the laser. Chemical lasers create the most powerful laser lights and are therefore the most dangerous lasers to handle. The U.S. Air Force wants to develop solid-state lasers that would be much lighter and safe enough to carry on jet fighters and maybe even tanks. This technology, however, based on the best possible estimates, is still twenty years away.

Thus, Colonel Moon's plot to destroy the land mines in the DMZ was quite impossible, despite such actions being shown in *Die Another Day*. If the producers had wanted to give the film a small margin of authenticity, they could have had Colonel Moon flying a 747 jet filled with massive chemical tanks used in tandem to create a powerful chemical laser. A plane armed with such a laser and flying only a few miles above the land mines could have blasted the mines away without much trouble. In reality, a scheme like that actually might have worked. It also would have explained why the colonel hadn't tried to just attack Seoul, and it would have provided plenty of thrills when the chemical tanks started to leak while the plane was attacked by U.S. jets. This proves once again that we should be writing scripts for James Bond movies instead of pointing out the serious inaccuracies in the science and the plots. We are available for reasonable rates if any intelligent film moguls are reading this chapter. Please contact us via e-mail.

||

Combating Germ Warfare

(and Other Nasty Matters)

We've discussed the grim possibility of the world being destroyed by nuclear winter, brought about by an atomic war started by SPECTRE. We've also analyzed superweapons of mass destruction such as *GoldenEye*'s gigantic space laser. So having discussed the first two Horsemen of the Apocalypse, let's turn to the others, Famine and Plague, or, as they are known today, biological warfare.

Virus Omega

If anything good could be said about the supervillain Ernst Stavro Blofeld, it was that he was very persistent. He never gave up, despite having his greatest triumphs turned into disasters by James Bond. In *Thunderball*, it was Blofeld's assistant, Emilio Largo, who died when SPECTRE's attempt to blackmail the United States and the United Kingdom with stolen atomic bombs failed. The stakes jumped when Blofeld worked for the mainland Chinese in *You Only Live Twice*. The plan was simple: start a nuclear war between the United States and the Soviet Union, with the Chinese taking control of the world after the war had ended. Nuclear winter most likely would have wiped out the Chinese even if Blofeld's scheme hadn't been stopped by 007. Undeterred, Blofeld refused to give up. He struck again, in *On Her Majesty's Secret Service*, the sixth Bond film (1969) and the only movie starring George Lazenby as our suave hero.

The virus Omega was a man-made plague designed by Blofeld that would cause infertility in plants, animals, and even humans. As proof of his skill in engineering animal viruses, Blofeld mentioned an outbreak of foot-and-mouth disease that had plagued England two years before (1967), implying that he was the one who had caused the disaster.

In this case, the screenwriters had done their homework. Foot-and-mouth disease had not been totally eradicated from Britain in the 1960s, but it was thought to be under control. A major outbreak of the virus in 1967, however, resulted in the slaughter of 196,000 cattle, 113,000 sheep, and 97,000 pigs. In all, 442,000 hoofed animals were killed to contain the disease, at a cost of approximately $750 million. Thus, Blofeld's threat to unleash a mutated, more deadly version of the virus on the animals of Britain and Europe would have been taken very seriously. In Thailand, for example, an outbreak of foot-and-mouth disease in 1997 resulted in the killing of 3.6 million pigs. A recurrence of foot-and-mouth disease in Great Britain in 2001 resulted in the slaughtering and burning of 7 million sheep and cows. The cost of the disease to the British economy was estimated at over $15 billion.

Foot-and-mouth disease is a highly infectious viral disease that's caused by an aphthovirus of the viral family *Picornaviridae*. The virus was first discovered in 1897. Foot-and-mouth disease afflicts animals with cloven hooves, such as cattle, pigs, and sheep. The disease begins with a fever, which is followed by blisters in the mouth and on the feet. Animals become lame and lose their appetite. They rapidly lose weight and produce less milk.

The disease is transmitted easily among animals through fluids such as blood, saliva, and milk. Fluid from broken blisters has especially high concentrations of the disease. It can be spread by people through shoes and clothing. Birds can transmit the disease, and dust particles from the soil have been known to carry the disease for hundreds of miles from the original infection point.

Foot-and-mouth disease kills only a small percentage of affected animals, primarily the young and old. While most animals recover from the virus, they are left weak and lame. A vaccine is available for the disease, but it is rarely used, because vaccinated animals, though free of the disease symptoms, can still carry the virus and pass it to other animals. Therefore, countries free of the disease refuse to import vaccinated animals. The only reliable method of stopping an outbreak of foot-and-mouth disease is killing all hoofed animals in the infected area and burning their bodies.

Foot-and-mouth disease is not dangerous to humans. During the 1967 British episode, one person was diagnosed with the disease, and that was an animal slaughterer who was exposed to vast amounts of tainted blood. Foot-and-mouth disease is entirely different from another viral infection called hand, foot, and mouth disease, which affects children. Humans eating meat from diseased animals aren't in any danger, as the virus that causes foot-and-mouth disease is eradicated by stomach acid.

As mentioned earlier, foot-and-mouth disease can have a drastic effect on a country's economy. When Britain's animals were infected in 2001, exhaustive precautions were taken to prevent the virus from spreading to Europe. Hundreds of British farms were closed. Sporting events such as horseracing, hunting, fishing, and rugby were canceled to reduce the role of human traffic in spreading the disease. Some schools closed for the same reason. Many national parks, zoos, and hiking trails were closed. At airports people were asked to disinfect their shoes. Numerous European countries, such as Germany, refused to allow any British meat across their borders.

This raises the question of how Blofeld planned to export his mutated virus to Britain or any other country. He had to rely on human frailty; in this case, allergies. In the Alps, Blofeld established a high-tech allergy clinic, where he cured women with terrible allergies. What these women didn't realize was that while curing them, Blofeld was also planting posthypnotic suggestions in their minds. Each woman was given a gift box containing a number of makeup accessories. The atomizer in the case contained the virus, while the compact mirror was a radio receiver. Blofeld's "angels of death" were conditioned to switch on the receiver to listen for his commands. Only a few sprays from the atomizer were necessary to destroy the British economy. If only James Bond hadn't destroyed Blofeld's operation and captured the women. But killing hoofed animals in England was the least of Blofeld's lethal threats.

Describing his genetic threats against Britain to Bond, Blofeld made it quite clear that the Omega Virus was designed to wipe out entire species. Blofeld threatened to cripple the balance of nature across the globe. It sounded extremely melodramatic and quite impossible. Unfortunately, it isn't as preposterous as it sounds.

In 1859, British colonists brought the first rabbits to Australia. Without natural enemies and with unchecked breeding, rabbits quickly became a major pest. In the early 1860s, red foxes were imported to Australia to kill the rabbits. Instead of wiping out the rabbit population, the wild foxes, with no natural predators to threaten them, multiplied in huge numbers and soon became an equally troubling problem.

In 1888, the famous scientist Louis Pasteur suggested killing off the rabbits by deliberately introducing a disease into their ecosystem. Approximately sixty years after his suggestion, the idea was actually tried.

Native Brazilian rabbits carry a harmless virus called myxomatosis, which is spread by mosquitoes. When European rabbits were exported to Brazil, it turned out that their immune systems could not deal with myxomatosis. Over 99 percent of the European rabbits imported to Brazil died from this disease. In 1950, the decision was made to introduce the Brazilian virus into rabbit warrens in remote sections of Australia in hopes of reducing the rabbit population.

Though this experiment was not conducted by a global villain like Ernst Stavro Blofeld, it was an actual example of biological warfare against an entire species. The disease spread across Australia, moving best when rainy conditions boosted mosquito growth. The 99 percent death rate held for some time, but finally the Australian rabbits adapted to the virus, as had the Brazilian rabbits. However, it was estimated that more than 80 percent of all rabbits in Australia died before the plague ended.[1] In the mid-1990s, it was estimated that there were 200 million rabbits in the Australian outback.

More than a half-century later, Australians are once again considering a virus to control a major animal pest. This time, the disease is a genetically engineered one, created by biotech engineers, and the animal is the red fox.

The introduction of the red fox in Australia was devastating to the native wildlife. Instead of wiping out the huge rabbit population, the fox attacked animals that had never seen it before and became easy prey to the carnivorous beast.

The fox is responsible for the extinction of dozens of small mammals on the isolated continent, and several species of other small mammals, endangered ground-nesting birds, amphibians, and reptiles are currently threatened. Equally troubling, the red fox carries rabies and can transmit the disease to humans. Rabies is currently unknown in Australia. However, the introduction of only a few animals infected with the disease to the ecosystem, along with the widespread range of the foxes across the continent, would be enough to cause disaster.

The virus currently being genetically engineered to eradicate the red fox involves a process known as immunocontraception. As implied by the long and formidable name, this technique turns the female red fox's immune system against male sperm. Thus, this virus renders female foxes infertile. Within a few generations, the fox population would drop to zero. Being a genetically engineered virus, the vaccine could be reworked if the foxes developed a resistance to the original strain.

Tests conducted outside Australia have proven immunocontraception an effective method of animal control. These tests were conducted in the United States to control the deer population. The virus was administered via injection, a method that is practicable only when the population group is small.[2]

For a large group like Australia's red foxes, the virus would have to be introduced gradually into the animal population using bait. Not everyone in Australia finds this idea attractive. Many scientists worry about the effect of the virus on other animals. Others are concerned that the targeted animals could adapt to the disease and become superfoxes, thus making them even more difficult to eradicate than they are already.[3] No one is entirely sure what effect a virus aimed at eliminating an entire species across a continent will have. It's a worry that has troubled many Australians for years and shows no signs of vanishing.

More troublesome is that what can be done with foxes, deer, and rabbits can be done with humans. Blofeld mentions to Bond that his Omega Virus could be modified to affect humans, if necessary. In a 1969 movie, such an idea sounded threatening but still unbelievable. More than thirty-six years later, with advances in genetic engineering, the same words sound quite possible—too possible. Creating a similar immunocontraception virus that would attack humans instead of animals would be a relatively simple and inexpensive biotech proposal. It's a project well within the capabilities of hundreds of biotech laboratories around the globe. The technology Ernst Stavro Blofeld bragged about is here. It's no longer just a threat in the movies; it's an actual threat to all life.

Poison from Space

In *Moonraker*, the eleventh James Bond movie, our hero battles the diabolical Hugo Drax, a multimillionaire industrialist who schemes to wipe out all human life on Earth by seeding the planet with a poison that affects only people. Afterward, Drax plans to repopulate the planet with his loyal employees, who are safe and protected on a hidden space station orbiting the globe.

This plot is so illogical that at times the film seems more parody than pastiche. The only thing the movie has in common with the original Ian Fleming book is the title, and even that is a stretch, accomplished only by calling space shuttles Moonrakers.

Drax doesn't lack ambition. Few men aspire to become gods. Like most megalomaniacs, the multimillionaire seems undaunted by consequences or doubts. Such an attitude might work when running a cut-throat business empire, but it's not the best business model when you're planning to manage an entire planet. Drax plans to kill over 4 billion people (the estimated world population in 1979). Mass murder on such a scale always leaves loose ends. No biological or chemical weapon is 100 percent efficient, but Drax remains unconcerned about survivors. Nor does he worry that his genetic pool might be too small to ensure the survival of the species. He's obviously a high-concept planner who doesn't believe in sweating the small stuff.

The linchpin of Drax's plan is a rare orchid that grows amid the ruins of an ancient civilization in the wilds of the Amazon rain forest. Exposure to the flower over a long period, Bond is told by the talkative Mr. Drax, turns people sterile. It's the cause of the decline and fall of the lost city. Drax's scientists have taken the orchid and amplified its powers, turning it into the world's deadliest airborne biological weapon. One sniff of the orchid dust causes instant death. Of course, this miracle toxin affects only humans, not plants or animals, thus leaving the world ready for human colonization.

There are two problems with Drax's plan. A killer poison that attacks only humans doesn't exist. Anything that kills people will usually kill other, similar mammals. So, in condemning the human race to death, Drax is also murdering much of humankind's food supply. That's not a good way to maintain the balance of nature.

The second problem with his tale is that there is no powerful poison connected with orchids. Wherever he got his information, it is wrong. There are plenty of beautiful orchids in the Amazon jungle, but none produces a toxin dangerous to humans. What he evidently meant, and should have said, is that his poison came from the castor-oil plant. The castor-oil plant flourishes in Brazil, which is one of the leading exporters of castor oil in the world. Also, the plant, with its bright red and white flowers, does produce in its seeds one of the world's deadliest poisons. The toxin is called ricin. Most likely, this is what Drax really planned to use to murder all humanity.

Ricin is the third most toxic substance known to man. The other two are botulism and plutonium. Contact with a crystal of ricin the size of a grain of salt causes death within forty-eight hours. The victim dies from convulsions, respiratory failure, and massive internal hemorrhaging as his or her blood cells explode.

Botulism can be destroyed by cooking. Ricin can't. The substance can be mixed in solids and liquids. It's also stable in aerosol form, so it can spread easily through the air. It is almost impossible to detect in the bloodstream, and there is no antidote or vaccine. It is more efficient than a small atomic bomb.

Terrorist groups around the world are starting to realize that ricin is the best, cheapest weapon of mass destruction. Formulas for processing it are available in mail-order handbooks from U.S.-based right-wing militia groups. Four or five dollars will buy enough castor beans to kill a dozen people.

In 1991, four members of the Minnesota Patriot's Council were charged in a scheme to put ricin on the doorknob and heater of a U.S. marshal's automobile. The men were apprehended with seven-tenths of a gram of ricin in their possession, which they had made in a home laboratory. They were sentenced to twenty months in jail.[4]

The one problem with ricin is that it's most effective when used in large amounts. For example, the amount of airborne ricin necessary to cover a 100-square-kilometer area and cause 50 percent lethality, assuming optimum dispersal conditions, is approximately 4 tons.[5] Since the world's surface consists of over 510 million square kilometers, the ricin needed to blanket the planet to achieve a 100 percent lethal dose is most likely more than Drax's space station could handle. Still, Drax had chemists working on the toxin, determined to increase its lethal power. Assuming they succeeded, perhaps the fifty containers of poison Drax intended to spread around the world would have sufficed to wipe out humanity. In 1979, such an idea seemed possible only in a James Bond movie. More than twenty-five years later, with ricin attacks on Capitol Hill and national security warnings of chemical warfare, it no longer seems improbable or unlikely. Proving again that even the worst Bond movies can contain a *seed* of truth.

12

Possible or Impossible?

They tell me that electricity's stored somewhere in here.
Science was never my strong point.
—SCARAMANGA TO 007, *THE MAN WITH THE GOLDEN GUN*

Possible or impossible? That's a question many moviegoers find themselves asking after watching the newest James Bond movie. Since the films are firmly set in the present and not the future, it seems logical to think that all of the science featured in the films would be based on reality, not fiction. Yet, as we've seen time and time again throughout this book, that's not always the case. Some of the most fascinating gadgets featured in the James Bond series are strictly products of a screenwriter's imagination, while other, seemingly impossible machines, exist today. In this chapter, we cover some of the devices not discussed elsewhere in this book that balance on the edge of possible or impossible.

Q's Robot Dog

This device played a minor role in *A View to a Kill*. Q called this prototype surveillance machine his "pet" because it looked like a dog. Actually, it closely resembled the robot dog K-9 in the British TV series *Dr. Who*. Q's machine was maneuvered by remote control and had a microphone and several cameras. Information was sent back to Q in his van. The weapons designer even used it to spy on Stacey Sutton in the shower.[1]

A very similar device, however, played a much more important role in real life after the terrorist attack on the World Trade Center on September 11, 2001. Called marsupial robots because they carried a smaller robot inside them for exploring tight spaces, dozens of these experimental surveillance robots, armed with bright lights and heat and motion detectors, crawled through the wreckage of the Twin Towers searching for survivors. Some of the machines were remotely controlled, while others were connected to the users by cables. The robots rode on caterpillar tracks and were powerful enough to push pieces of concrete out of their way.[2] We can't help but think they would have made Q very proud!

The Minijet

Octopussy, the thirteenth Bond movie, included a remarkable gadget that seemed impossible at the time: a miniature jet, about a dozen feet long, powered by a single jet engine.

At the beginning of the film, James Bond is discovered trying to destroy a radar plane in Cuba. Bond escapes to his car, which has a horse trailer in tow. Inside the trailer is the miniature jet, and off Bond flies. His enemies refuse to let him get away, so they fire a heat-seeking missile after him. Some deft maneuvering through a hangar saves Bond, and the scene ends with the missile destroying the radar plane.

Possible or impossible? The minijet was possible and was actually used in filming the sequence described. Several of its flying tricks were clearly not possible, however, and were made believable only by special effects. As with many items supposedly designed by Q Division, the jet was a one-of-a-kind item.

The plane was built and piloted by the stunt flier Corky Fornof. The minijet was called Bede Acrostar Mini Jet. The 12-foot-long plane, which weighed only 450 pounds, was the smallest operational jet in the world in 1983. Powered by a microturbo TRS-18 single-jet engine, the Acrostar had a climb rate of 2,800 feet per minute. It could carry only one passenger and was capable of flying over 300 mph, with a ceiling of 30,000 feet. After its movie appearance, the jet toured U.S. air shows for years as the Coors Light Microjet and, later, as the Bud Light Microjet.

The Minibreather

With so much of the story in *Thunderball* taking place underwater, it was necessary for Q to come up with an easy-to-use, compact underwater breathing mechanism for James Bond. This minibreather, as Q called it, held about four minutes' worth of oxygen. It consisted of a small mouthpiece with a valve to start the supply of oxygen flowing. Attached to the valve were two small tanks that held the oxygen. These tanks were so small that each was about the size of half a cigar.

In the film, Bond used the minibreather twice. The first instance was when a speedboat cut his standard scuba tanks. He used it a second time when he escaped from a pool of hungry sharks.[3]

Unfortunately, the minibreather is impossible as shown. The smallest underwater oxygen tank available is an emergency air container about the size of a soda can. It's known as the Spare Air system and is designed to get someone to the surface during an emergency. There are three Spare Air models, all at a pressure of 3,000 psi. The 170 model holds 1.7 cubic feet of air or approximately thirty breaths at the surface, and the 300 and 300-N Nitrox models both hold 3 cubic feet of air or approximately fifty-seven breaths at the surface. It's estimated that the 170 model has enough air to breathe for approximately two minutes.[4]

Solex Agitator

In *The Man with the Golden Gun*, Bond battles Francisco Scaramanga, the world's most deadly assassin. A freelance killer who needs only one bullet to kill his targets, Scaramanga has stolen the powerful Solex Agitator for one of his clients. The device is a solar energy–focusing machine that's proclaimed to be the solution to the world's energy crisis. Being no fool, Scaramanga murders his client and keeps the Agitator for himself. The villain intends to auction off the device to the highest-bidding nation. In the meantime, he has installed the only actual Agitator on his island base, which is run entirely by solar energy. The powerful energy converter also doubles as the power pack for a deadly laser cannon. Although 007 manages to defeat Scaramanga in a deadly cat-and-mouse game, he's unable to save Scaramanga's hideout from the ineptitude of his partner, Mary Goodnight. Bond and Goodnight just manage to escape before the island explodes. But they rescue the Solex Agitator nonetheless, leaving us to wonder where all the cheap solar energy is.

According to the film, the Solex Agitator is a 95 percent efficient solar energy converter. On Scaramanga's private island, thermoelectric generators convert solar energy into electricity. Everything is fully automated, and only one employee is needed to run the entire plant. When Scaramanga says, "They tell me the electricity's stored somewhere in here. Science was never my strong point," Bond answers, "Superconductivity coils cooled by liquid helium. If I were you, I wouldn't stick my finger or anything else over the rail. At four hundred fifty-three degrees below zero Fahrenheit, that liquid helium would break it up like an icicle."

Scaramanga's basic operation is simple. When he presses a button in his power plant, a hidden door in the mountainside hideout opens and a giant reflector mirror comes out of the rock. The mirror opens into a parabolic dish, the dish locks onto the sun, and the sun's energy flames into the Agitator. The Agitator gathers the solar heat and transmits it to the thermogenerators.

Bond, who is superscientific in this movie, says, "Reflecting through this, that panel must generate three and a half thousand degrees Fahrenheit." To which Scaramanga replies, "We can focus the power wherever we want." A big device goes from the Solex into a focus machine, which looks like a gigantic ray gun. When Scaramanga presses a button, the ray gun device blows up the airplane Bond used to fly to the secret island base. It's yet another example of a screenplay that equates lasers and solar energy.

Ignoring all the mumbo-jumbo pseudoscientific remarks exchanged between Bond and Scaramanga, we do know that solar energy is real and is considered an important power source for the future, as was the case back in 1974, when *The Man with the Golden Gun* first appeared. Still, most of the world runs on oil, not on solar energy. As we asked earlier, where are the Solex Agitators?

Unfortunately, the Solex Agitator exists only in the imagination of the movie's screenwriters. While the basic concept of harnessing solar energy to provide cheap, clean energy is true, the method is far more complicated and much less efficient than shown. But we are moving in the right direction, if somewhat slowly.

Solar energy is solar radiation that reaches the Earth from the sun. It can be converted directly or indirectly into other forms of energy, such as heat or electricity. The two major problems with dealing with solar energy are the variable way it reaches the Earth and the large amount of area needed to collect enough energy for it to be useful. At present, solar energy is used to heat water and buildings and to generate electricity.

Electric plants are working hard to develop photovoltaics, a process where solar energy is directly converted into electricity. At present, most electricity produced by solar energy comes indirectly from steam generators that use solar thermal collectors to heat a working fluid, as shown in *The Man with the Golden Gun*.

Photovoltaic (PV) energy converts sunlight into electricity through the use of a PV cell, also known as a solar cell. These cells are nonmechanical and usually are made from silicon alloys.

Sunlight is composed of photons. The photons contain various amounts of energy corresponding to the different wavelengths of the solar spectrum. When photons strike a PV cell, they may be reflected, pass right through, or be absorbed. The absorbed photons are the ones that provide energy to generate electricity. When enough sunlight is absorbed by the material, electrons are dislodged from the cell's atoms. Special treatment of the material surface during manufacturing makes the front surface of the cell more receptive to free electrons, so the electrons naturally migrate to the surface.

When these electrons leave their position, holes are formed. When many electrons, each carrying a negative charge, travel toward the front surface of the cell, the resulting imbalance of charge between the cell's front and back surfaces creates a voltage potential like the negative and positive terminals of a battery. When the two surfaces are connected through an external load, electricity flows.

The PV cell is the basic building block of a PV system. Individual cells can vary in size from about half an inch to about 4 inches across. One cell produces only about 1 or 2 watts, which isn't much power. To increase output, cells are electrically connected into a packaged module. Then modules are connected to form an array. The term *array* refers to the entire generating plant, whether it is made up of one or several thousand modules. As many modules as needed can be connected to form the array to any specified size.

Obviously, the performance of a PV array is dependent on sunlight (the first problem mentioned). Climate conditions have a major effect on the amount of solar energy received by a PV array and, therefore, how well it generates electricity. At present, PV modules are about 10 percent efficient in converting sunlight to electricity. Scientists hope to raise this efficiency to 20 percent in the near future. Compare these figures to the 95 percent efficiency conversion accomplished by the Solex Agitator and it becomes clear why every major country in the world wanted the device so badly.[5]

The PV cell was invented in 1954 by Bell Telephone researchers examining the sensitivity of a specially prepared silicon wafer to sunlight. Beginning in the late 1950s, PVs were used to power U.S. space satellites. The success of PVs in space generated commercial applications for PV technology. The simplest PV systems power many of the small calculators and wristwatches used every day.[6]

The most advanced solar power sources concentrate solar

energy by using a parabolic trough. This system has a linear parabolic-shaped reflector that focuses the sun's radiation on a linear receiver located at the focus of the parabola. The collector tracks the sun along one axis from east to west during the day to ensure that the receiver is continuously focused on the sun. Due to its parabolic shape, a trough can focus the sun at 30 to 100 times its normal intensity on the receiver, achieving operating temperatures over 752 degrees Fahrenheit. Not the 3,500 degrees Scaramanga bragged about, but then, the Solex Agitator worked with 95 percent efficiency.

Thus, when asked if the Solex Agitator is possible or impossible, we're forced to answer: impossible now, but, we hope, possible someday in the future.

The Wetbike

In *The Spy Who Loved Me*, the U.S. Navy is given orders to blow up Atlantis, the gigantic floating base of the villain Karl Stromberg. Stromberg is holding Major Anya Amasova, a beautiful Russian agent, as his prisoner. James Bond persuades the submarine captain to hold off with his torpedoes for an hour so he can travel to Atlantis to rescue his female counterpart. He does so by riding on a water motorcycle sled called a Wetbike. This one-person water bike made a big impression on the audience, as few had ever seen a Wetbike before.

In the movie, Bond assembled the bike on the submarine from a bunch of parts sent to him in a large sack by Q. Roger Moore actually rode the device in the movie, getting very wet in the process!

The original Suzuki-powered, 50-horsepower version of the Wetbike was introduced in 1978 by Spirit Marine, a division of Arctic Enterprises.[7] Early advertisements marketed the product as the Wetbike Watercycle: a motorcycle on water with a combination of all the thrills and fun of motorcycling, boating, and waterskiing.[8] Throughout its production years, from 1978 through 1992, the Wetbike design went through various changes, upgrades, technical advancements, and safety modifications. The most notable changes were the substitution in 1985 of a strong, exceptionally lightweight Metton body unit, and in 1986 the introduction of a 60-horsepower, 800cc Suzuki motor. The new motor raised the performance of the Wetbike enough so that it was rated the fastest personal watercraft available at that time. According to the Ian Fleming Foundation Web site, the Wetbike ridden by James Bond in *The Spy Who Loved Me* "is the actual engineering prototype of the first Wetbike ever built. The Wetbike as built by Nelson Tyler was donated to the Ian Fleming Foundation in the fall of 1992."[9]

The Automatic Targeting Attack Communicator

In *For Your Eyes Only*, James Bond is sent to recover the Automatic Targeting Attack Communicator (ATAC) encryption device from a British warship that has sunk off the coast of Albania. The Russians are also trying to obtain the device, which will enable them to control Polaris nuclear submarines. The British ATAC system seems realistic. It uses an ultralow-frequency coded transmitter to order submarines to launch ballistic missiles. In today's world, computer systems abound that issue directions to submarines and missile-launching systems. So the major technical aspect of this Bond film is plausible and most likely possible, yet is in no way remarkable.

The 3-D Visual Identigraph

Another plausible but unremarkable device in *For Your Eyes Only* is the film's 3-D Visual Identigraph, which Q is using in its experimental stage. Suitably for 1981, Bond installs two huge tapes that are as big as old turntable records. The tapes have photos of all the villains collected from the entire world: Britain, the United States, Germany, France, and so on. To find a villain in the collection, Bond describes the person he wants to identify, and then Q programs the information. The drawing appears on a computer screen.

A pretty woman serves coffee to Q and Bond—wait a minute, what happened to the politically correct, new and improved Bond? At any rate, the Identigraph provides a computer printout of a photo that matches the person Bond is seeking. This type of technology is used routinely by today's police departments and is seen on numerous TV shows such as *CSI: Crime Scene Investigation*, which means everyone uses it. But it was cool for 1981.

13

Some Thoughts about Secret Bases

The firing power inside my crater is enough to annihilate a
small army. You can watch it all on TV. It's the last
program you're likely to see.
—Ernst Stavro Blofeld to 007, *You Only Live Twice*

As we remarked earlier in this book, great crimes need great criminals. Great criminals, of course, are filled with great ambitions. In most cases, such men (and women) need a vast number of confederates to handle all the busy work of committing the momentous crime. The execution of evil doing requires large spaces, plenty of weapons, and lots of supplies. And, of course, the bigger the crime, the more complex the requirements, which inevitably leads to a secret criminal base of operations.

Just as James Bond wouldn't be Bond without his scientific gadgets, amazing wristwatches, and death-defying automobiles, Ernst Stavro Blofeld, Hugo Drax, Gustav Graves, and other megavillains wouldn't be anywhere near as threatening without their gigantic headquarters inside volcanoes, on invisible space stations, or in palaces of ice. Everything in the Bond universe is big and bold; criminals never direct their schemes from back rooms in seedy hotels. Only the biggest and brightest hideouts serve as proper shelter for these megalomaniac monsters. Whatever mad goal they seek, these villains pursue it in style.

As we've seen in previous chapters, however, style doesn't always convert into reality or even possibility. Are the gigantic hideaways used by Bond's greatest enemies possible? Or, equally important, affordable?

Blofeld's Volcano

In *You Only Live Twice*, Ernst Stavro Blofeld is working for the Chinese communists, trying to start a nuclear war between the Soviets and the Americans. To accomplish his aim, Blofeld is kidnapping space stations deployed by the two countries and placing the blame on the other nation for the crime. A reasonable person might ask why Blofeld is kidnapping the astronauts and secretly holding them hostage when killing them would be so much less expensive. That same reasonable person might also ask why U.S. or Soviet officials would believe the other superpower is kidnapping their men. What real use is there for kidnapped astronauts? Asking such questions evidently violates the same rule of moviegoing that requires the audience not to wonder why the Chinese communists would want to start a nuclear war and most likely vaporize all life on Earth. And why would a brilliant criminal like Blofeld agree to a plan that would most likely result in his inability to spend his reward? Basic logic doesn't work when analyzing big-budget spy flicks.

No matter. Ignoring any sort of logical thinking, we note that Blofeld's giant rocket base is hidden in an inactive volcano in a remote section of Japan. The mouth of the volcano is covered by a huge metal roof that is disguised to look like a lake. Whenever Blofeld wants to launch his spaceship-eating rocket, the roof splits open, the rocket takes off, and the roof closes. As the base is located in a very remote section of Japan with only primitive fishing villages located nearby, no one seems to notice the giant multistage rocket flying overhead spewing gas and flames over the ocean. Obviously, these people have endured too many Godzilla attacks to be frightened by something so minor as a spaceship.

Before tackling the obvious question of how Blofeld managed to build a secret base inside an inactive volcano, we need to raise an even more pressing question: how did Blofeld *find* an inactive volcano in which to build his base? Did he place an advertisement looking for "volcanic property" in all the Japanese newspapers? Or could he have found the volcano by chance, stumbling across an ad for "one inactive volcano, for sale, cheap"? The question might sound ridiculous, but we do have to wonder how, in an age of rampant population growth, criminal masterminds seem to be able to gobble up all the property they need without anyone complaining about price fixing.

Building a huge space base inside the walls of an extinct volcano must have required hundreds of workers and technicians months of effort, which had to be undertaken in the strictest secrecy so no word of their mission got back to the Japanese government. We can only assume that Blofeld employed a large fleet of helicopters to bring his workers into the volcano, and that once an employee signed on with SPECTRE, his contract bound him to the company for life. The only way to maintain the secrecy of the hidden base was to make sure no one ever left the volcano.

The base includes a helicopter pad; the helicopter serves as the main transport method in and out of the hidden base. Blofeld, we learn, employed Osato, of Osato Chemicals, to deliver liquid oxygen to the hideout to fuel his rocket ship. We can only assume that Osato delivered basic supplies for the base personnel at the same time.

Along with a helicopter pad, the secret base also contains a massive concrete rocket launching pad for Blofeld's spaceship, along with all the necessary equipment for fueling the ship, putting the rocket in the right position for launch, and diverting the flames on takeoff. Huge metal shutters protect Blofeld's launch control from the rocket's flames.

Also worth mentioning is the monorail system that travels throughout Blofeld's domain (there are living quarters, laboratories, storerooms, and such built into the lower depths of the volcano). And, in strict observance of the madmen's code of villainy, Blofeld keeps a pool of hungry piranha near his lounge chair in case he needs to discipline any of his less competent employees.

Probably the most expensive item in Blofeld's base is the giant reusable spaceship with its gigantic claw mouth that is used to entrap space capsules. In the late 1960s, a rocket this size must have run in the neighborhood of $500 million to build, and that's without the extras installed by Blofeld's technicians, such as the clawlike mouth and the ship's ability to land standing straight up instead of on its side. Such a rocket ship was well beyond the technology of any nation of the time, and even today there are no rockets that are able to land in this fashion. Blofeld probably could have sold his discovery legally to the U.S. space program and not have bothered with the volcano.

It's safe to conclude that Blofeld's volcano base cost in 1960s money well over a billion dollars to build, maybe even more, depending on whether the SPECTRE agents were unionized. The reward of $100 million for starting a war between the United States and the Soviet Union was an extremely bad deal for Blofeld. He should have held out for being named ruler of Australia, or, at least, ruler of Japan.

The Moonraker Space Station

If we conclude, based on the previous description, that Blofeld's volcano base was the most outrageous criminal headquarters on Earth, then Hugo Drax's space station qualifies as the most startling hideaway in the entire Bond universe. According to the end credits of *Moonraker*, the science adviser for the film was the well-known science writer Eric Burgess. We can only assume from watching the film that most of Burgess's advice was ignored, or terribly misunderstood.

In *Moonraker*, the multimillionaire industrialist Hugo Drax plans, from his space station circling the globe, to wipe out Earth's population. Drax's hideaway is invisible from Earth because of radar-jamming equipment on the station. Exactly how this equipment, invented long before stealth technology was ever imagined, works is never explained. Nor does anyone bother to explain how the space station, even if it was invisible to radar, would also not be seen by people using telescopes. It's difficult to imagine a space station some 200 yards in diameter going completely unnoticed as it orbits Earth.

The lack of intelligent thought that went into the design and deployment of this invisible space station is staggering. How, one wonders, was this huge station assembled in space, in secret? How did Drax keep the voyages made by his fleet of Moonraker spaceships hidden, since none of them used the radar-jamming equipment on the space station? More important, how did Drax's team of spaceworkers construct a huge orbital space station in just a few years, when, in forty years of space travel, the United States and Russia haven't been able to build a station one-tenth the size of the Drax base?

When the fleet of Moonraker spaceships docks at the invisible space station, one person switches the gigantic base back on. Evidently, whenever Drax's people leave the station, they turn off all the lights—and the power—and the artificial gravity. When the power, lights, and gravity are switched back on, the person flipping the switches reports through a microphone that the "life support system is nominal."

Gravity is a force that acts between two masses. It is not something that we can create artificially. What we can do on a space station is create a force that feels like gravity. Centripetal force is the force that acts on an object to keep it in circular motion. This force is directed to the center of rotation. Centrifugal force is not a force at all; it's merely the lack of centripetal force.

In other words, if we rotate a space station, we generate centripetal force on objects inside the station. This force causes these objects to "fall" to the outside wall of the station. This centripetal force feels like gravity. It is experienced as a centrifugal force on the floor (wall) beneath one's feet. As the outside wall of the station would feel like "down," the center of the station would feel like "up." Life in an artificial gravity field would be quite different from anything shown in science-fiction movies or TV shows, as the usual meanings of "up" and "down" are very different in outer space.

The force of the artificial gravity felt by the inhabitants of the space station would equal the centripetal acceleration of the station. This force is defined by the formula $a = v^2/r$, where a is the acceleration of the station, v is the rotational velocity, and r is the radius. Thus, the faster the station spins, the greater the acceleration felt by our space station crew. The larger the radius of the station, the smaller the acceleration. Since the radius of a particular station always remains the same, it follows that by changing the rotational velocity of the station, we could actually simulate the gravity on Earth.

Drax's space station rotates in space and thus has artificial gravity. It's not the up and down as defined by centripetal force, but that's a forgivable sin, considering the movie budgets and special effects of the time. Not so forgivable is the one scene that relies on artificial gravity for its kick.

In perhaps the silliest scene ever to appear in a James Bond film, our hero and Dr. Holly Goodhead are captured by Jaws on the space station and brought to Drax in the station's huge control room. A U.S. space shuttle filled with space marines is on its way to the space station. The space station is equipped with a giant laser cannon, however, and Drax plans to use it to destroy the shuttle. There's no way to stop the villain, not even when Jaws suddenly decides to help Bond.

Our hero seems doomed until he suddenly spots a black button on a nearby control panel that's labeled "Do not use unless station secured. Emergency Stop." An emergency stop button on a space station? The button is only a few feet away. Bond's hands are free and everyone's attention is on the approaching shuttle, so 007 leaps forward and pushes the button. Instantly, a retro rocket fires, the space station comes to a complete stop, artificial gravity vanishes, and Drax is flung far away from the space laser controls. The U.S.

space shuttle opens its hold and the space marines attack. The world as we know it is saved thanks to the fast thinking of James Bond, or at least to his fast button pushing.

We are left to wonder what possible reason there is to have an emergency stop button on a gigantic space station. It's a question we suspect even 007 can't answer.

14

Famous Last Lines

This concludes our book about the science of James Bond. But, like most writers, we can't resist a few additional comments about the contents and concepts that we've discussed in this work.

The number-one question we're asked whenever we discuss our science books is: why do you write them? A few people seem to think that we hate the subjects we cover, that we despise James Bond and find comic books unreadable, and that we never watch Japanese animated shows. Actually, we're both huge fans of everything we discuss in our books. We wouldn't write about James Bond and the movies featuring him if we didn't love the films and the characters. Criticism does not mean contempt. All we do in our books is examine the science in the stories and discuss whether the material is believable or not. We criticize what we find lacking, just as we praise the material we find stimulating. The prime purpose of criticism is to encourage creators to do better. We love the James Bond films. All we ask is that they use logical story lines and intelligent science.

The secret of writing an entertaining James Bond novel or producing a successful James Bond movie is to create a suspension of disbelief. That is, while we are reading or watching, the story seems believable and not terribly outrageous. That's not easy to do when constructing a spy adventure where the hero drives a car with an ejector seat and the bad guy is a genetically altered Nazi superman. Such stories work only if the rest of the work remains firmly anchored in the everyday world. In other words, everything else in the adventure needs to stay accurate to real life, including the science. Without some believable science, the James Bond movies would fly off into the realms of fantasy. We hope that our comments and corrections and even suggestions help the films stay firmly rooted to the earth.

In re-viewing all the Bond movies (numerous times) while writing this book, we each came away with certain lines of sparkling dialogue that struck us as the best of 007. We included a number of these quotes at the beginning of each chapter, but obviously with so many great lines, we couldn't include them all. So we each chose a favorite exchange to conclude this final chapter.

Bob's choice features 007's first meeting with Ernst Stavro Blofeld.

> Blofeld: James Bond. Allow me to introduce myself. I am Ernst Stavro Blofeld. They told me you were assassinated in Hong Kong.
> Bond: Yes, this is my second life.
> Blofeld: You only live twice, Mr. Bond.

Lois chooses an opening exchange in her favorite Bond movie, *Dr. No.*

> Bond: I admire your courage, Miss . . . ?
> Sylvia Trench: Trench. Sylvia Trench. I admire your luck, Mr. . . . ?
> Bond: Bond. James Bond.

And now, as you ponder your favorite Bond lines, Bob and Lois are off to find a good martini—shaken, of course, not stirred.

APPENDIXES

APPENDIX A

THE BOND BOOKS

The sales of Bond books total more than 50 million copies and are still going strong.

Novels and Short Stories

By Ian Fleming

All books are listed with the original publication date as well as the publisher and date of the most recent reissue.

Casino Royale (1953), New York: Penguin Books (September 2002)

Live and Let Die (1954), New York: Penguin Books (May 2003)

Moonraker (1955), New York: Penguin Books (December 2002)

Diamonds Are Forever (1956), New York: Penguin Books (December 2002)

From Russia with Love (1957), New York: Penguin Books (December 2002)

Doctor No (1958), New York: Penguin Books (September 2002)

Goldfinger (1959), New York: Penguin Books (September 2002)

For Your Eyes Only (1960) (short stories), New York: Penguin Books (May 2003)

Thunderball (1961), New York: Penguin Books (May 2003)

The Spy Who Loved Me (1962), New York: Penguin Books (September 2003)

On Her Majesty's Secret Service (1963), New York: Penguin Books (September 2003)

You Only Live Twice (1964), New York: Penguin Books
   (September 2003)
The Man with the Golden Gun (1965), New York: Penguin Books
   (April 2004)
Octopussy and the Living Daylights (1966), New York: Penguin
   Books (April 2004)

Fleming also wrote a series of short stories featuring James Bond,
some of which bear the titles of Bond films:

"Quantum of Solace" (1959), *Cosmopolitan Magazine*, May
"For Your Eyes Only" (1960), first published in the short story
   collection *For Your Eyes Only*
"From a View to a Kill" (1960), an original story, first published
   in the short story collection *For Your Eyes Only*
"The Hildebrand Rarity" (1960), *Playboy Magazine*, March
"Risico" (1960), an original story, first published in the short
   story collection *For Your Eyes Only*
"The Living Daylights" (1962), *Sunday Times Colour Section*
   (London), February 4
"The Property of a Lady" (1963), *The Ivory Hammer: The Year at
   Sotheby's*, 219th Season, 1962–1963 (London: Longmans)
"007 in New York" (1964), in Ian Fleming, *Thrilling Cities* (New
   York: New American Library)
"Octopussy" (1966), two-part serialization in *Playboy*, March
   and April

By Kingsley Amis, writing as Robert Markham

Colonel Sun (1968), New York: HarperCollins (February 1993)

By John Gardner

Licence Renewed (1981), New York: Jove Books (July 1987)
For Special Services (1982), New York: Berkley Publishing Group
   (July 1987)
Icebreaker (1983), New York: Jove Books (July 1987)
Role of Honour (1984), New York: Putnam (September 1984)

Nobody Lives Forever (1986), New York: Berkley Publishing Group (February 1990)

No Deals, Mr. Bond (1987), New York: Berkley Publishing Group (September 1991)

Scorpius (1988), Diamond Books (Baltimore: December 1989)

Win, Lose, or Die (1989), New York: Berkley Publishing Group (August 1990)

Brokenclaw (1990), New York: Berkley Publishing Group (March 1991)

The Man from Barbarossa (1991), New York: Berkley Publishing Group (January 1992)

Death Is Forever (1992), New York: Berkley Publishing Group (August 1993)

Never Send Flowers (1993), New York: Berkley Publishing Group (June 1994)

SeaFire (1994), New York: Berkley Publishing Group (June 1995)

Cold Fall (1996), New York: Berkley Publishing Group (July 1997)

By Raymond Benson

Zero Minus Ten (1997), New York: Jove Books (August 1998)

The Facts of Death (1998), New York: Jove Books (August 1999)

High Time to Kill (1999), New York: Jove Books (June 2000)

Doubleshot (2000), New York: Jove Books (June 2001)

Never Dream of Dying (2001), New York: Jove Books (April 2002)

The Man with the Red Tattoo (2002), New York: Jove Books (April 2003)

Raymond Benson also wrote three short stories featuring James Bond:

"Blast from the Past," *Playboy*, January 1997

"Midsummer Night's Dream," *Playboy*, January 1999

"Live at Five," *TV Guide*, November 13 and 19, 1999

Film Tie-Ins

By Christopher Wood

> *James Bond: The Spy Who Loved Me* (1977), New York:
>   HarperCollins
> *James Bond and Moonraker* (1979), London: Cape

By John Gardner

> *Licence to Kill* (1989), Charter Books
> *GoldenEye* (1995), New York: Berkley Publishing Group

By Raymond Benson

> *Tomorrow Never Dies* (1997), New York: Berkley Publishing
>   Group
> *The World Is Not Enough* (1999), New York: Berkley Publishing
>   Group
> *Die Another Day* (2002), New York: Berkley Boulevard Books

APPENDIX B

THE OO SECRET AGENTS

Although there is some historical basis for the 00 designation, we assume—perhaps incorrectly, as it is never made explicitly clear in the novels or films—that a secret agent receives the 00 designation only after he or she kills two people. Somehow, this fact then gives the agent a "license to kill."

This means that 007 is not the first 00 agent. In fact, there may be six such agents in operation, numbered 001 through 006, and an unknown number with designations of 008 and up. We have to review the novels and movies to answer the question: How many 00 agents are in the British Secret Service with James Bond?

In Ian Fleming's novel *Moonraker*, Bond is the senior 00 agent of three such agents. The other two agents are 008 (Bill in Berlin) and 0011 (missing in Singapore). The three agents share an office. Later, in the novel *Goldfinger*, 008 is mentioned again, this time as a replacement if Bond is killed. In Fleming's novel *Thunderball*, 009 assumes leadership of the 00 section while Bond is on sick leave. In the novel *On Her Majesty's Secret Service*, 006 is described as a former Royal Marine commando.

So the 00 tally from Ian Fleming's novels is five: 006, 007, 008, 009, 0011.

Subsequent writers of the Bond novels created other 00 agents. In Robert Markham's (aka Kingsley Amis) *Colonel Sun*, we have 005, or Stuart Thomas, whose eye defect forced him to leave the 00 unit and become head of Station G in Greece.

Agent 005 returns in Raymond Benson's *Facts of Death*, along with 004, whom Bond replaces as the investigator of the murder of M's lover.

In Benson's novelization of *The World Is Not Enough*, 0012 is the agent who is killed while trying to retrieve Sir Robert King's money. So, using the novels as our guide, we have eight: 004, 005, 006, 007, 008, 009, 0011, and 0012.

In the movies, we find more extensive use of 00 agents, although we must note that the list of agents is not particularly long. In *The Man with the Golden Gun*, Bill Fairbanks operates as 002 until Francisco Scaramanga kills him in Beirut, Lebanon. Another 002 agent replaces Fairbanks in *The Living Daylights*. In this case, the 002 title is not given during the movie, but it is displayed in the closing credits. Next up is 003, who is killed in Siberia during *A View to a Kill*. Agent 003 appears in the film as a corpse. *The Living Daylights* also includes 004, who is murdered by a KGB assassin in Gibraltar. *GoldenEye* features 006, or Alec Trevelyan, who supposedly dies during an attack on the Arkhangelsk chemical warfare facility, but who later appears as head of Janus.

Although 008 is never on screen in *Goldfinger*, he is mentioned several times. When M fears that Bond may try to avenge the murder of Jill Masterson, he threatens to replace 007 with 008. When Bond believes that he's going to be killed by Auric Goldfinger's laser beam, he says that 008 will take his place.

In *The Living Daylights*, M again threatens to replace Bond with 008 if Bond refuses to kill General Leonid Pushkin. Although 008 never appears on screen, he is the world's best replacement agent. If your boss wants to fire you from your secret agent job, he or she brings in 008 to take over.

Octopussy gives us 009, who is killed by Mischka and Grischka in East Berlin. 009 is the man dressed as a clown in the opening scene of the film. *In The World Is Not Enough*, 009 returns to kill the terrorist Renard, who kidnapped Sir Robert King's daughter.

In *Thunderball*, nine chairs are set up in an MI6 conference room, where all the 00 agents in the world will meet. We are led to believe that there are only nine 00 agents worldwide. Thus, based on the movies, we have seven: 002, 003, 004, 006, 007, 008, and 009.

Combining our lists, we end up with ten 00 agents: 2 to 9, 11, and 12. As we just mentioned, however, when all the 00 agents are brought in from the field in *Thunderball*, there are only nine chairs. Does that mean there is an extra agent? Most likely not.

While all the 00 agents lead dangerous lives, they rarely seem to work together, except for in *The World Is Not Enough*. Thus, the chances of several 00 agents dying at the same time seems remote. Assuming that is the case, then to avoid any confusion between agents, nine spies would need ten codes. With ten codes available, one number would always be unused. Thus, whenever a 00 agent is killed, his or her replacement would be given the unused number, making it clear to other agents in the field that he or she is a new agent. The dead agent's number would remain unused until another 00 is killed. We assume that the number 010 is kept out of circulation because of its resemblance to binary code. As to 001, we can only assume that the creator of Bond, Ian Fleming, reserved that number for himself.

APPENDIX C

THE BOND CARS

Films that aren't mentioned didn't feature a car of any note.

Dr. No   At the beginning of the film, Bond drove a light blue 1961 Sunbeam Alpine convertible, series 5.

From Russia with Love   Early in the film, Bond drove a 4.5 liter Bentley Sports Tourer, the favorite car of Ian Fleming.

Goldfinger   Q gave Bond a silver metallic Aston Martin DB5 with special equipment, including two machine guns that moved forward from behind the parking lights. The car also had a rear smokescreen, and, when needed, a steel plate rose from the trunk as a bullet-proof shield. The license plate could be rotated electronically with plates for Great Britain, France, and Switzerland. Tire cutters emerged from the car's hubcaps. The car was also equipped with a passenger ejector seat with a removable roof panel. A mapping screen was located behind the radio and tracked a homing device. There was also an oil spray behind the right rear turn signal and a tube that ejected three-pointed nails onto the road.

Thunderball   Bond drove an Aston Martin DB5. It was equipped with twin pipes beneath the rear bumper that poured water on gangsters. The Aston Martin DB5 made brief appearances in *GoldenEye* (1995) and *Tomorrow Never Dies* (1997) and was treated as Bond's personal car.

You Only Live Twice   This movie featured a Toyota 2000 GT used by the Japanese Secret Service. The car contained a closed-circuit TV system with cameras behind the license plates that recorded everything that could be seen in front of or behind it. It also carried a cordless telephone, a hi-fi receiver, a cassette player that turned itself on as soon as voices could be heard, a video recorder in the glove compartment, and a miniature color TV.

The Spy Who Loved Me   Agent 007 used a Lotus Esprit with rear-firing cement tubes. When driven into the ocean, the car turned into a two-person miniature submarine. It was equipped with surface-to-air missiles, depth charges, an underwater smoke screen, and torpedoes.

For Your Eyes Only   Bond drove a Lotus Esprit Turbo, equipped with a unique locking device. When thieves tried to break into the car, it exploded.

The Living Daylights   Bond switched to an Aston Martin DBS, equipped with diverse extras such as rocket drive, steel spikes, and a device for cutting through ice.

GoldenEye   This film featured a BMW Z3, the first time Bond drove a German car. He drove it only a short time in Cuba before handing it over to his American contact.

Tomorrow Never Dies   In this film, Bond rented a BMW 750 in Hamburg, Germany. The car was fully loaded with the usual Bond car weapons. Additionally, it could be operated by a hand-held remote control. In a desperate chase through a parking garage, Bond steers the car from the backseat, and later controls the auto even when he's not inside it.

The World Is Not Enough   Bond drove a BMW Z8 sports car, which featured a complete remote control unit, an onboard navigation system, and a hidden rocket system in the car's side vents.

Die Another Day   Bond's car was an Aston Martin V12 Vanquish. It was an exceptionally well-equipped car, with four grille-mounted rockets as well as two machine guns. There were two motion-detecting guns concealed beneath the hood and spiked tires for driving on ice. The car's most astonishing feature was a form of invisibility camouflage.

APPENDIX D

THE SCIENCE OF MARTINIS

In chapter 7 of the first Bond book, *Casino Royale* (1953), James Bond tells a bartender how to make his favorite drink. According to Bond, the drink should be made in a deep champagne goblet and consist of "three measures of Gordon's, one of vodka, half a measure of Kina Lillet. Shake it very well until it's ice-cold, then add a large thin slice of lemon-peel. Got it?"

Kina Lillet is a type of vermouth. Gordon's is a brand of gin. Where most martinis are made with gin, and vodka martinis are made with just vodka, Bond mixes them in one drink. He calls this concoction a vesper, the name of the female double agent in the novel.

By shaking the martini, you make the drink cold, which is key for most vodka drinks to taste good. It also mixes the vodka and gin nicely, giving the drink a smooth texture, though we must point out that not all martini experts agree about the shaking.

Numerous bartenders have pointed out that a classic martini is stirred, not shaken, because shaking bruises the gin. Gin is made with juniper berries, which are very delicate. Bruised gin tastes, at least according to martini aficionados, like rubbing alcohol. Martini experts say that shaking a martini makes it too sharp.

There are three differences between a martini (or a vodka martini) that has been stirred and one that has been shaken. First, a shaken martini is colder than one stirred, since the ice has had a chance to swirl around the drink more. Second, shaking a martini dissolves air into the mix; this is the notorious "bruising" of the gin that makes a martini taste too sharp. Third, a shaken martini will more completely dissolve the vermouth, giving a less oily feel to the drink.

In 1999, Professors Maurice Hirst and John Trevinthick from the Department of Biochemistry at the University of Western Ontario in Canada decided to run tests on Bond's martinis. The two professors know that a little alcohol is good for you and reduces your risk of having cardiovascular disease, strokes, and cataracts. Needless to say, James Bond, Secret Agent 007, suffers from none of these problems.

The two scientists focused on the antioxidant actions of the alcohol in the drinks. That led them to check whether there was a difference in the antioxidant activities of shaken and stirred martinis.

Oxidants in some parts of a human body might be bad. Antioxidants in the vegetables you eat are good. However, antioxidants bought in a bottle are most likely not good for anyone's health.

The two professors made mini-martinis using 6 milliliters of gin mixed with 3 milliliters of vermouth. These drinks were then shaken or stirred. The scientists then decided to measure how good a martini was at breaking down oxidants. The easiest way to do this was to measure the effect different martinis had on hydrogen peroxide, which is a byproduct of all the martinis.

Gin by itself left 58 percent of the hydrogen peroxide behind— not very useful. Vermouth, however, left only 1.9 percent. Stirred martinis had even better results, removing all but 0.157 percent of the hydrogen peroxide. But the winners of the study were the shaken martinis, which left only 0.072 percent of the hydrogen peroxide behind.

The conclusion reached by the two professors was published in a letter in the Christmas issue of the *British Medical Journal* in 1999. The two professors wrote, "007's profound state of health may be due, at least in part, to compliant bartenders."

The article provoked a number of outraged comments from scientists around the world disputing the claim that a vesper was a healthy drink. Some writers argued about taste, some argued about chemistry, and a few even argued that the effects of cocktail olives in the drink needed to be addressed.

The health benefits of the Bond martini made news all around the world, once again demonstrating the power and popularity of James Bond. So the next time you're planning to raise a toast to 007, make sure it's a vesper and that the drink is shaken, not stirred. You'll live longer.

Notes

1. Uncovering the Origins of Mr. Bond: Spies and Science

1. *Wikipedia: The Free Encyclopedia*, http://en.wikipedia.org/wiki/Richard_Sorge.
2. The specific James Bond martini recipe is: Put 4 tablespoons of the best vodka into a silver cocktail shaker; add 1 tablespoon of exceptionally dry vermouth; add ice and shake—make certain that you do not stir, as shaking makes the martini colder and cloudy (that is, mysterious); add one green olive; add one thin slice of lemon peel. For the true James Bond experience, wear a tuxedo or dinner jacket with black tie and get someone to serve you the martini on a silver tray.

2. Sending Secret Messages: Superspy Decoder Rings

1. The Codes and Ciphers Web site, www.codesandciphers.org.uk/enigma, and the University of Salerno, www.dia.unisa.it/professori/ads/corso-security/www/CORSO-9900/crittografiaclassica/home.us.net/~encore/Enigma/enigma.html.
2. Matt Blaze's Cryptography Resource, www.crypto.com; "The Clipper Chip," Electronic Privacy Information Center, www.epic.org/crypto/clipper/; and Tony Sale, "Codes and Ciphers in the Second World War," www.codesandciphers.org.uk/.
3. SKIPJACK and KEA Algorithms, http://csrc.nist.gov/CryptoToolkit/skipjack/skipjack-kea.htm, and Cryptography, http://world.std.com/~franl/crypto.html.
4. Paul Todd and Jonathan Bloch, *Global Intelligence: The World's Secret Services Today* (New York: Zed Books, 2003), 50.
5. Ibid.

3. Building a Bond Car: Flying, Underwater, and Missile-Launching Aston Martins

1. Kevin Bonsor, "How Flying Cars Will Work," http://travel.how stuffworks.com/flying-car1.htm.
2. Ibid.
3. Ibid.
4. Museum of Flight, "Collections," www.museumofflight.org/Display.asp?Page=Collections.
5. Bionic Dolphin, www.bionicdolphin.com/cms.
6. "The Bionic Dolphin," *Discover Magazine*, September 1992.
7. Rinspeed Web site, www.rinspeed.com/pages/content/frames_e.htm.
8. Gibbs Technologies Web site, www.aquada.co.uk.
9. Auto Command Professional Series Remote Car Starter, www.gizmocity.com/starter.html.
10. "Optical Camouflage," http://projects.star.t.u-tokyo.ac.jp/projects/MEDIA/xv/oc.html.

4. Arming Yourself (and Other Explosive Ideas)

1. The Nobel Foundation, www.nobel.se/nobel/alfred-nobel/biograph ical/life-work/nitrodyn.html.

5. Stopping Nuclear War

1. "Avro Blue Steel Standoff Missile," www.skomer.u-net.com/pro jects/bluesteel.htm.
2. U.S. Census Bureau statistics as of 2005.
3. Alan F. Phillips, "The Immediate Effect: Medical Problems Are Nuclear War," *3 A.M.* (2001), www.3ammagazine.com/magazine/issue_5/articles/nuclear_war_medical_effects.html.
4. Ibid.
5. Ibid.

6. Using Your Senses: Assorted Body Equipment

1. John Cork, *James Bond, the Legacy: Forty Years of 007 Movies* (New York: Harry N. Abrams, 2002).
2. SpeechStudio Web site, www.speechstudio.com/commercial/speech.htm.
3. Comp.speech Frequently Asked Questions Web site, http://fife.speech.cs.cmu.edu/comp.speech/.

4. Soft14.com Web site, www.soft14.com/Utilities_and_Hardware/Security_and_Encryption/Virtual_Personality_4396_Review.html.

5. "Fingerprints and Trace Evidence," North Carolina Wesleyan College, http://faculty.ncwc.edu/toconnor/315/315lect05.htm.

6. Ibid.

7. Ibid.

8. BBC News, "Doubt Cast on Fingerprint Security," May 17, 2002, http://news.bbc.co.uk/1/hi/sci/tech/1991517.stm.

9. Ibid.

10. Ann Harrison, "Hackers Claim New Fingerprint Biometric Attack," *SecurityFocus*, August 13, 2003.

11. Ibid.

7. Getting Away from It All: In the Air and on the Sea

1. Ryan Dilley, "Guns, Goldfinger and Sky Marshals," *BBC News Online*, May 20, 2003, http://news.bbc.co.uk./1/hi/uk/3039583.stm.

2. Dean Speir and Mark Moritz, "Debunking 'The Goldfinger Myth,'" the Gun Zone (2000–2004), www.thegunzone.com/091101/goldfinger.html.

3. Robert J. Horsky, United Airlines Cabin Crew, O'Hare Airport, in an interview conducted at O'Hare Airport, United Airlines Terminal, by Robert Weinberg on May 1, 2004.

4. Ibid.

5. "Autogyros," www.jefflewis.net/autogyros.html (January 4, 2002).

6. John Meyer, "Hydrofoil Design Basics—A Brief Tutorial," (November 11, 2000), www.foils.org/basics.htm.

7. Leslie Hayward, "The History of Hydrofoils," *Hovering Craft and Hydrofoils* 4, no. 8 (May 1965) to 6, no. 6 (February 1967).

8. United States Hydrofoil Association, "Hydrofoiling History," www.hydrofoil.org/history.html.

9. "Hovercraft and Hydrofoils," the K–8 Aeronautics Internet Textbook, http://wings.avkids.com/Book/Vehicles/advanced/hovercraft-01.html.

10. "Hydrofoil," *Wikipedia: The Free Encyclopedia*, http://en.wikipedia.org/wiki/Hydrofoil.

8. Getting Farther Away from It All: Outer-Space Shenanigans

1. Marshall Brain, "How Compasses Work," http://science.howstuffworks.com/compass3.htm.

2. See our book *The Science of Superheroes*, in which we explain all about rems and radiation as it relates to Bruce Banner and the Hulk.

3. NASA Facts, "Robert H. Goddard: American Rocket Pioneer, www.gsfc.nasa.gov/gsfc/service/gallery/fact_sheets/general/goddard/goddard.htm, and "Robert Goddard and His Rockets," www-istp.gsfc.nasa.gov/stargaze/Sgoddard.htm.

4. "Rocket Equations," http://my.execpc.com/~culp/rockets/rckt_eqn.html and "Questions and Answers about Flying and Mission Control," http://quest.arc.nasa.gov/space/ask/flying/Rocket_Speed.txt.

5. Marshall Brain, "How Rocket Engines Work," www.howstuffworks.com/rocket.htm.

6. Dr. Marc Rayman's Mission Log, http://nmp.jpl.nasa.gov/ds1/.

7. Tier One Private Manned Space Program, www.scaled.com/projects/tierone/index.htm, and Richard Seaman, "First Flight of SpaceShipOne into Space," www.richard-seaman.com/Aircraft/AirShows/SpaceShipOne2004.

8. Maggie McKee, "Beefed-Up Motor Will Boost SpaceShipOne," www.newscientist.com/news/news.jsp?id=ns99996396.

9. Tier One Private Manned Space Program, www.scaled.com/projects/tierone/message.htm and www.scaled.com/projects/tierone/faq.htm.

10. "Moonraker a Reaction to Star Wars?, Not So Fast," Commander Bond.net.forums (July 21, 2004).

11. Roger D. Launius, "Sputnik and the Origins of the Space Age," www.hq.nasa.gov/office/pao/History/sputnik/sputorig.html.

12. Anatoly Zak, "The True Story of Laika the Dog," www.space.com/news/laika_anniversary_991103.html.

13. History of the Space Shuttle, www.hq.nasa.gov/office/pao/History/shuttlehistory.html.

14. "President Bush Announces New Vision for Space Exploration Program," www.whitehouse.gov/news/releases/2004/01/20040114-1.html.

15. National Aeronautics and Space Administration, http://spaceflight.nasa.gov/shuttle/reference/factsheets/asseltrn.html.

10. Nullifying the Threat of Superweapons

1. Major Scott W. Merkle, "Non-Nuclear EMP: Automating the Military May Prove a Real Threat," www.fas.org/irp/agency/army/miph/1997-1/merkle.htm.

2. Ibid.
3. Ibid.
4. As reported in "Stop That Car or We'll Zap You!" *Micro-Wave News*, September/October 1996, 1.
5. Dan Baum, "Electronic Armageddon," *Wired*, March 2002, 2.
6. George Smith, "Chupacabras of Infowar," *Crypt Newsletter*, July 1997, 3.
7. Ibid.
8. Paul Eng, "E-Bombs Could Spell Digital Doomsday," ABCnews.com, October 9, 2001, 2.
9. Jim Wilson, "E-Bomb," *Popular Mechanics* (September 2001), 3.
10. Ibid.
11. George Smith, "The Chupacabras of Infowar," *Crypt Newsletter*, March 23, 1999, 4.
12. Ibid., 5.
13. Brian Hsu, "Army Slams Press for Scaring People with Bogus Weapon," *Taipei Times*, September 18, 2002, http://taipeitimes.com/news/2002/09/18/story/0000168470.
14. George Smith, "The Chupacabras of Infowar," *Crypt Newsletter*, July 22, 1997, 1.
15. "New Equipment Brings Higher Output, Not Higher Costs," *Electronic News*, October 4, 1999.
16. "The Science of James Bond: The Villains' Master Plans," *BBC News*, November 18, 2004, www.bbc.co.uk/science/hottopics/jamesbond/villains.shtml.
17. Ibid.
18. "Loma Prieta Earthquake" (August 28, 2004), http://en.wikipedia.org/wiki/Loma_Prieta_earthquake.
19. "Compaq Computer Corporation Hosts FEMA Earthquake Risk Assessment Meeting in Silicon Valley," Cupertino, California, September 22, 1999, www.fema.gov/regions/ix/1999/r9n19.shtm.

11. Combating Germ Warfare (and Other Nasty Matters)

1. "Green Goo—Life in the Era of Humane Genocide," www.homoexcelsior.com.
2. Stewart Taggart, "Taming Australia's Wild Kingdom," *Wired News*, September 10, 2002, 1.
3. Ibid, 2.
4. Grover C. Bailey, "Ricin, A Deadly Toxin That Targets Police," *Dimension News Online*, http://dimensionsnews.com/index.htm.

5. Ferdinando L. Mirachi and Michael Allswede, "CBRNE Ricin," *E-Medicine*, June 29, 2004, www.emedicine.com/emerg/topic889 .htm.

12. Possible or Impossible?

1. "The Science of James Bond: 007's Gadgets," *BBC News*, November 18, 2004, www.bbc.co.uk/science/hottopics/jamesbond/bondgadgets .shtml.
2. Ibid.
3. "The Science of James Bond: 007's Gadgets: Underwater Breather," *BBC News*, November 18, 2004, www.bbc.co.uk/science/hottopics/ jamesbond/bondgadgets.shtml.
4. Spare Air, "Frequently Asked Questions," 2003–2004, www.spareair .com/service/.
5. "Solar Energy," *Energy Information Administration*, June, 4, 2004, www.eia.doe.gov/kids/energyfacts/sources/renewable/solar.html.
6. Ibid.
7. Wetbike Web site, www.wetbike.net (January 2002).
8. Ibid.
9. *Goldeneye Magazine*, no. 1 (Fall 1992), 15.

Index